REAL MEN
WORK WITH WORDS

CROSSWORDS FOR MEN

Fill in Books with 100 Puzzles

PUZZLE THERAPIST

CROSSWORD | SUDOKU | KIDS & ADULTS

CONTENTS

PUZZLE 1

ACROSS

1. Voice mail prompt

5. Wry comic Mort

9. Fibula neighbor

14. Mideast's Gulf of ___

15. Shall I compare ___ to a summer's day?'

16. Fill in the blank with this word: ""I Still See ___" ("Paint Your Wagon" tune)"

17. Fill in the blank with this word: "___ occasion (never)"

18. Fill in the blank with this word: "Astronomy's ___ cloud"

19. Fill in the blank with this word: "___ latte"

20. "Dance the night away!"

23. Harvesting for fodder

24. It's in the back row, right of center

25. Tea time, perhaps

27. Swindler

32. Old Greek coins

36. Fill in the blank with this word: ""Morning Dance" band Spyro ___"

38. Rubaiyat' rhyme scheme

39. Snowflake or crystal shape

41. Susan who co-starred in "Five Easy Pieces"

43. Research facility: Abbr.

44. Oven ___

46. Std. on food labels

47. Triple-platinum 1982 album with the #1 hit "Africa"

49. Fill in the blank with this word: ""___ tale's best for winter": Shak."

51. Fill in the blank with this word: "___ tide"

53. Recovers, with "up"

58. "So?"

63. Talk show host Lake

64. The English translation for the french word: omettre

65. Fill in the blank with this word: "Catch ___"

66. 'Twenty Years After' character

67. Three-part ordeal for H.S. students

68. Fill in the blank with this word: "Auvers-sur-___, last home of Vincent van Gogh"

69. Over here...'

70. Explosives

71. Vintage vehicles

DOWN

1. Was emboldened

2. Fill in the blank with this word: "___ vincit amor"

3. Sluggo's comics pal

4. Mushroom variety

5. Like some traffic

6. Yes, matey! Sailors use this word to hail a ship, or to attract attention

7. a heron rookery

8. Stop working so hard

9. Some of Moby's music

10. Zoological wings

11. Fill in the blank with this word: ""That's ___ ""

12. Org. for the Denver Gold and Chicago Blitz

13. Writer's supply: Abbr.

21. Out of shape

22. Puerto ___

26. Fill in the blank with this word: "___ fruit"

28. Snoozes

29. Picasso's muse Dora ___

30. Opening run

31. Okinawa port

32. What ___?'

33. The center of the Czech Republic's wool industry, it looks like it needs to buy a vowel

34. Where hops are dried

35. Time-share unit

37. Word of disappointment

40. W.W. II-era G.I., e.g.

42. Slangy street greeting

45. Surveyor's assistant

48. Refuse to yield

50. With the situation thus

52. Fill in the blank with this word: ""___, of golden daffodils": Wordsworth"

54. Tryster's escape route, maybe

55. Fill in the blank with this word: "___, meenie, miney, mo"

56. Fill in the blank with this word: "___ Rizzo, Dustin Hoffman role"

57. Tart fruits

58. Fill in the blank with this word: ""That's a ___!""

59. Website statistic

60. Worms cries

61. W.B.A. stats

62. verb secure with a bitt

PUZZLE 2

ACROSS

1. Canio's wife in "Pagliacci"

6. Fill in the blank with this word: "___ Beach, Hawaii"

9. Antiaircraft missile

14. Fill in the blank with this word: "1997 film "___ Gold""

15. You're in balance if you know it's the complement of "yang"

16. Swing and a miss

17. Nobelist Bohr

18. She, in S

19. Quiche ingr

20. Popular cable network ... or a hint to 17-, 24-, 47- and 56-Across

23. Fill in the blank with this word: ""How sweet ___!""

24. "It has come to my attention ..."

25. Sue Grafton's '___ for Innocent'

28. Fill in the blank with this word: "___-Deutschland"

29. Work ___ sweat

30. Like some port authorities

32. The right to have a jury of these, meaning one's equals, not lords, is mentioned in the Magna Carta

34. Fill in the blank with this word: ""Last one ___ a rotten egg!""

35. Suffered serious consequences

41. Guy Lombardo's "___ Lonely Trail"

42. Sci-fi case

43. Hare ___ (religious sect)

47. Wray of "King Kong"

48. Department of eastern France

51. Finish this popular saying: "You are what you___."

52. It operates by compression

54. Sandy tract, in Britain

55. "Eeeww!"

58. Shoot

60. The English translation for the french word: point par pouce

61. Fill in the blank with this word: "___ Island National Monument"

62. Fill in the blank with this word: "As ___ (usually)"

63. When Guy Fawkes Day is celebrated: Abbr.

64. Fill in the blank with this word: "___ vie"

65. Fastenable, as labels

66. Suffix with social

67. When cutting these flowers, use a spring clothespin on the stems to keep from getting pricked

DOWN

1. Vatican emissary

2. Writer known as Old Possum, and his family

3. Fill in the blank with this word: ""Would you not ___ breathed?": Shak."

4. Takes out

5. Made a tax valuation: Abbr.

6. Where you may need to read the fine print

7. Fill in the blank with this word: "___ Coyote"

8. Word before republic or seat

9. They might be crossed

10. TV's "___-Team"

11. Time's 2001 Person of the Year

12. Worst grade

13. Baseball positions: Abbr.

21. Holy mackerel!'

22. TNT part

26. Suffix with phosphor

27. Zaire's Mobutu ___ Seko

29. Pupil's place

31. Wear down

32. Some palmtops, for short

33. Vodka, peach schnapps, cranberry juice & grapefruit juice make up one version of this "on the beach"

35. Toll road

36. Gillette ___ Plus

37. For real?'

38. The English translation for the french word: effusif

39. Word with roll or bar

40. Yarn strand

44. Fill in the blank with this word: "___-Dazs"

45. Tiny criticism

46. 2002 Katherine Frank political biography

48. Wind god

49. wrong (similar term)

50. Promontories

53. Subject of a 1982 best seller

54. 7 Faces of ___' (1964 film)

56. Hawaii's ___ Bay

57. Will who played Grandpa Walton

58. Was on the bottom?

59. Part of C.P.I.

PUZZLE 3

ACROSS

1. Sinatra staple

6. Usurp

11. Without further ___

14. You look like you've ___ ghost!'

15. Fill in the blank with this word: "___ Sorrel (woman in a love triangle in "Adam Bede")"

16. Rock's ___ Jovi

17. King's honor

20. Personal air

21. McDonald's mascot before Ronald

22. Sweetie

26. Fill in the blank with this word: ""___ Heartbeat" (Amy Grant hit)"

27. Suffix with symptom

28. Magician's name ending

31. Oxford bottom

32. Seabiscuit jockey ___ Pollard

33. Young starlet's promoter, maybe

37. The English translation for the french word: psychothÈrapeute

42. Semicircular windows

43. Worker in a garden

44. Welfare, with "the"

47. Secure online protocol

48. The English translation for the french word: saisir

49. Several Norwegian kings

51. Tree that yields fragrant balsam

55. The plural of the word addendum

58. Town outside of Buffalo

59. House wear

64. Long-running B'way musical seen by couples?

65. Type of parallelogram

66. Wigwam relative

67. Saison d'___

68. With time to spare

69. Fill in the blank with this word: "Belgian violin virtuoso Eugene ___"

DOWN

1. Fill in the blank with this word: "___ Explorer (Web browser)"

2. USN enlistee

3. The English translation for the french word: toile

4. Weak

5. Supporter of the mascot Handsome Dan

6. Fill in the blank with this word: ""___ Fall in Love" (1961 hit by the Lettermen)"

7. Old New Yorker cartoonist Gardner ___

8. Truncation indications: Abbr.

9. One ___ at a time

10. Novel subtitled "A Peep at Polynesian Life"

11. Waiter

12. Ones out for a while

13. Jack of spades feature

18. Zip

19. Overhaul

22. Wrists, anatomically

23. Sport ___ (modern vehicles)

24. The English translation for the french word: ordonner

25. Some musical notes

29. The English translation for the french word: Tanguy

30. Nagy of Hungary

33. Moo ___ pork

34. Zillions

35. Suffixes with sultan

36. Western treaty grp.

38. Witty

39. Fill in the blank with this word: ""___ your pardon?""

40. Former N.F.L. great Junior ___

41. Fill in the blank with this word: "___ limits (election issue)"

44. Within reach

45. Fill in the blank with this word: ""The ___ trick in the book!""

46. Sonny boy

48. Smallest bone in the human body

50. Zzzz

52. Young role on TV

53. Ransom ___ Olds

54. void (similar term)

56. Qatar's capital

57. Omnia vincit ___

60. T-shirt sizes, in short

61. Water tester: Abbr.

62. Lana Del ___, singer with the 2014 #1 album 'Ultraviolence'

63. GM: "___ the USA in your Chevrolet"

PUZZLE 4

ACROSS

1. Laid bets at a casino

6. Young eel

10. This __ laughing matter!'

14. Eyes

15. Fill in the blank with this word: "___-tiller"

16. Call in the game Battleship

17. The English translation for the french word: ThaÔs

18. What's ___ for me?'

19. The Soup ___

20. What kind, decent people wear?

23. Work unit

24. Fill in the blank with this word: "___ a kind (pair)"

25. Takeaway game

26. Peace Nobelist Kim ___ Jung

29. Bad place for the modest

32. Symphonie espagnole' composer

35. Fill in the blank with this word: ""Little ___, you're really lookin' fine" (1964 lyric)"

36. Discovered

37. Fill in the blank with this word: ""Thanks ___!""

38. Israel's Barak and Olmert

41. Spittoon sound

42. Paint job finale

44. Fill in the blank with this word: "Chinese author ___ Yutang"

45. Salinger's 'For ___ - With Love and Squalor'

46. Actress Garr discovered at a statue site?

50. Wichita-to-Omaha dir.

51. Pitcher Robb ___

52. When repeated, a Thor Heyerdahl title

53. Fill in the blank with this word: "___ Pictures (old studio)"

56. Darned

60. ___ latte

62. Pooch's name

63. The English translation for the french word: complÈter

64. Iodized _____ lick

65. "Little" barnyard bird with an alliterative name in a classic Willie Dixon blues song

66. Fill in the blank with this word: "Cost ___ and a leg"

67. Full of compassion

68. Aloud

69. With 74-Down, leaving shortly

DOWN

1. Perplexed pronouncement

2. Liniment user

3. Fill in the blank with this word: "___ Thai (official name of Thailand)"

4. Fill in the blank with this word: ""We'll give a long cheer for ___ men" ("Down the Field" lyric)"

5. "The Simpsons" character who often refers to himself in the third person

6. D. W. ___ Award, honor for 58-/46-Across for lifetime achievement

7. Writer Jaffe

8. Fill in the blank with this word: "Don Cornell classic, "___ Fair""

9. Wuthering Heights' genre

10. Fill in the blank with this word: ""___ idea!""

11. Stars

12. Pince-___

13. Maritime CIA

21. Fill in the blank with this word: "___-podge"

22. Fill in the blank with this word: ""___ the opinion that ...""

27. Per ___ (yearly)

28. Vocalist Gorme

29. Fill in the blank with this word: "___ home (out)"

30. People

31. Ran

32. Grow dark

33. Fill in the blank with this word: "___ Highway (route through Whitehorse, Yukon)"

34. It's up for grabs on a court

39. Given to procrastination

40. Winding

43. Fill in the blank with this word: "___ penny (very common, in British lingo)"

47. Brad of "Sleepers"

48. More black

49. Like some Egyptian relics

53. Morocco's capital

54. Small ridge on the edge of a button or dial

55. Vermont's ___ Mountain Resort

57. Printing technique: Abbr.

58. Work of prose or poetry

59. Middle of a run?

60. Union-busting grp.?

61. Lyricist ___ David

PUZZLE 5

ACROSS

1. There was a prescription for war in 1839 when Britain & China went toe-to-toe over this drug's trade

6. What knows the drill, for short?

10. United ___ Emirates

14. Dnieper tributary

15. Switch suffix

16. Tiny complaints

17. "... and he's got Budweiser and Michelob on tap - excellent ___!"

19. Fill in the blank with this word: "Christie's "Death on the ___""

20. The Sound of Music' song

21. Meetings of the minds?

23. H

25. Unctuous

26. Turing test participant

29. The English translation for the french word: nuque

31. Alligator's kin

35. Work without ___

36. Turns sharply

38. Big bill

39. It's set in a castle near Seville

43. Regional woodland

44. Popular caramel candy

45. Fill in the blank with this word: "Farmer's ___"

46. Though

48. 1940s USSR secret police

50. Worrying sound to a balloonist

51. 1960 Updike novel

53. Withered

55. Tracking these, coast guard rates them as bergy bits, growlers, small, medium, & large

59. Kind of smoothie

63. Film producer ___ Al-Fayed

64. Succumbs to interrogation, perhaps

66. Kind of romance between actors

67. Peer group?

68. The English translation for the french word: tinter

69. Singer Lovett

70. What Texas hold'em tables hold

71. Native of the central Caucasus

DOWN

1. Weirdo

2. Spanish "but"

3. Munich's river

4. Starving

5. Wrestler

6. Retailer with stylized mountaintops in its logo

7. Tolkien creatures

8. Fill in the blank with this word: "___ Rose"

9. The English translation for the french word: cosmique

10. Pesky

11. New York's Jacob ___ Park

12. Mythical king of the Huns

13. Popular vacation locale

18. Flair

22. Spanish man's name that means "peaceful"

24. Peeping Tom

26. Wood used to make surfboards

27. First black major-league baseball coach Buck ___

28. Wt. of some flour sacks

30. Thomas Hardy's ___ Heath

32. Zoo dividers

33. Trac II competition

34. Some Dodges

37. Pimlico garb

40. Job for an orthodontist

41. New ___ (Congolese money)

42. Old-fashioned ones

47. Rutabaga, e.g.

49. Something to bust

52. Writer Marsh

54. French peers

55. TV's "American ___"

56. Brit's teapot cover

57. Fill in the blank with this word: ""James Joyce" author Leon ___"

58. Struck, once

60. Thurman et al.

61. Oscar-winning French film director ___ Cl

62. U.S.A.F. NCO

65. Sue Grafton's '___ for Evidence'

PUZZLE 6

ACROSS

1. Key related to F# minor: Abbr.

5. Yum-Yum, Peep-Bo and Pitti-Sing in "The Mikado"

10. Fill in the blank with this word: ""48___""

13. Starbucks stores

15. Fill in the blank with this word: "By land ___"

16. Strauss's "___ Heldenleben"

17. Pair of pants?

19. Traveler's alternative to J.F.K.

20. Sidi ___, Morocco

21. More like a cold shower?

23. Fill in the blank with this word: "___ Digital Shorts (late-night comic bits)"

25. The English translation for the french word: noyau actif de galaxie

27. Trapped

28. Where ___ sign?'

29. Yes, matey! Sailors use this word to hail a ship, or to attract attention

31. Rap's Dr. ___

32. Vardalos and Peeples

34. Unfreeze

36. Wedding reception party?

40. Seminal mystery of 1887

43. Spotted ___

44. Explosives

45. Yemen's capital

46. Fill in the blank with this word: "___ and snee"

48. Whit

50. Fill in the blank with this word: ""No ___""

51. Want in the worst way

55. Fill in the blank with this word: "___ de plume"

56. Tournament sit-out

57. Like some chords

59. Red, as a Spanish wine

61. Small island

62. Mystery desserts?

66. Long-running B'way musical seen by couples?

67. French poet (born in Romania) who was one of the cofounders of the dada movement (1896-1963)

68. Pope's "___ Solitude"

69. W.W. II vessel: Abbr.

70. Pulitzer-winning journalist Seymour

71. African fox

DOWN

1. Fill in the blank with this word: ""___ du lieber!'"

2. Fill in the blank with this word: "Egyptian ___ (cat breed)"

3. Station that's part of a TV network

4. One of the acting Bridges

5. The ___ Squad' of TV and film

6. The sculptures "Cloud Shepherd" and "Coquille Crystals"

7. Pancreatic hormone

8. Shield

9. Refuge near a battle

10. Sun: Prefix

11. The English translation for the french word: Rigel

12. Web

14. Yemen's capital

18. Bedwear: Var.

22. One's wife, slangily

23. -

24. whirring (similar term)

26. successful (similar term)

30. Single-named New Age musician

33. Strip in Hollywood

35. Actor Sam

37. Beach cookouts

38. Fill in the blank with this word: "___-weensy"

39. Story, in France

41. William ___, who founded Ralston Purina

42. Where some errands are run

47. Charge, to a physicist

49. Shell alternative

51. The Beatles' "Any Time ___"

52. Street magician ___ Angel

53. Words to the maestro

54. What stealth planes avoid

58. Ones in charge: Abbr.

60. Fill in the blank with this word: "Actress ___ Pinkett Smith"

63. Fill in the blank with this word: "___-di-dah"

64. The English translation for the french word: …os

65. 1983 Indy winner Tom

PUZZLE 7

ACROSS

1. "High Hopes" lyricist

5. Western scenery

10. Harem rooms

14. What were the names of the 3 Cartwright sons?

15. Mother ___

16. Femme fatale in "The Carpetbaggers"

17. Old soap, perhaps

19. Yearbook sect.

20. What's missing from a KO?

21. Willowy: Var.

22. The English translation for the french word: humble

23. Mother of Xerxes I

25. Tiny fraction of a min.

27. Bragging sort

33. Witty Nash

36. U.S.A.F. rank

37. Fill in the blank with this word: ""___ Haw""

38. Those, to Robert Burns

39. Vast extents

40. Unscramble this word: lgri

41. Fill in the blank with this word: ""There is no ___ team""

42. Cube-hopping character in a 1980s arcade game

43. Swindler's victim

44. *Stumbled upon

47. Part of a Spanish play

48. Illegal firings

52. Batting helmet feature

54. Verdi's work on this General's life opens in Cyprus; Shakespeare's tale begins in Venice

58. You might not be able to stand this

59. Fill in the blank with this word: "Aglio e ___ (pasta dressing)"

60. beautiful handwriting

62. Wild goose

63. Veldt sight

64. Uris's "___ 18"

65. Had an unquiet sleep

66. Spanish soccer star Sergio ___

67. Fill in the blank with this word: ""The even mead, that ___ brought sweetly forth ...": "Henry V""

DOWN

1. In an infamous "Cheers" episode, this character's husband, Eddie LeBec, was run over by a Zamboni

2. Unlikely to reconsider

3. Fill in the blank with this word: ""I ___ idea!""

4. Name placeholder in govt. records

5. In the center

6. Rabbit ___

7. Women of Andaluc

8. Vapory beginning

9. They clean locks

10. Start of a cry by Juliet

11. Bull's-eye

12. Seraph of S

13. Mont. neighbor

18. TV's ___ twins

24. Go on a vacation tour

26. T-shirt sizes, in short

28. Topps rival

29. Wing: Prefix

30. Sales slips: Abbr.

31. Suffixes with ballad and command

32. Unscramble this word: leyr

33. Fill in the blank with this word: "___ suspension (ear medication)"

34. Karmann ___ (old Volkswagen)

35. Dolphin leader

39. Maker of a dramatic 1971 getaway

40. Wanders aimlessly

42. Yemeni shrub whose leaves are chewed as a stimulant

43. Victim of a 1955 coup

45. Parrot

46. Some lunches

49. Biblical land with "ivory and apes and peacocks"

50. Tough as ___

51. Time-honored Irish cleric, for short

52. Years on end

53. Fill in the blank with this word: "Chrysler Building architect William Van ___"

55. Anatomical tissue

56. City in Judah

57. Middle of a run?

61. The English translation for the french word: PME

PUZZLE 8

ACROSS

1. Fill in the blank with this word: "___ Falls"

6. Masked critter

10. Vulnerable gap

14. Water bearers

15. Western Indian

16. Fill in the blank with this word: ""Forever, ___" (1996 humor book)"

17. What a poor diet may need

20. Paris's ___-de-Medecine

21. Broadcast

22. The English translation for the french word: DBA

24. Fill in the blank with this word: "Barbara Kingsolver's

"___ Am"

27. Yeshiva leader

28. Brand for hay fever sufferers

31. 1950s fad item

33. On the ___

34. Unscramble this word: tnuaer

36. Microsoft chief, to some

38. Pirate flag in the summer sun?

42. Bake, as eggs

43. Winner of all four grand slam titles

45. U.S. Army E-7

48. Wicked Game' singer Chris

50. Texas has one, in song

51. Natalia Makarova joined this company upon graduation from the Leningrad Choreographic School

53. Fill in the blank with this word: "Allan ___, "Sands of Iwo Jima" director"

55. They're not part of the body: Abbr.

56. Actress Massey et al.

58. ___ bar (popular candy)

61. See 17-Across

66. Fill in the blank with this word: "___ way, shape or form"

67. one of five children born at the same time from the same pregnancy

68. Uncommon?

69. Fill in the blank with this word: "Eye ___"

70. Fill in the blank with this word: "___ Genesis (old video game console)"

71. The Joads, e.g.

DOWN

1. Velocity meas.

2. Island off India's coast

3. Interest

4. Jazz buff

5. Fill in the blank with this word: "___ Fjord"

6. Fill in the blank with this word: ""The Iceman ___""

7. Taxonomic suffix

8. Suffix with ball or bass

9. Tennis miss

10. Stiff hairs

11. Coffee orders

12. The name of this one-celled protozoan comes from the Greek for "change"

13. Comic Howie

18. Roy Wood's band before Wizzard

19. Study of the atmosphere

22. Rock's Steely ___

23. Yawn-inducing

25. Chafes

26. Fill in the blank with this word: ""___ volat propriis"

(Oregon's motto)"

29. Tracks

30. of or relating to or contained in or serving as an archive

32. Terra ___ (pulverized gypsum)

35. Twin sister of Ares

37. Watchdog's sound

39. toward the mouth or oral region

40. Internal passageways

41. Sound of a leak

44. What "y" might become

45. Never mind!'

46. Department store founder who pioneered credit unions

47. Tries out for 'American Idol,' maybe

49. V.I.P.

52. Stoned, in a way

54. Pitcher Robb ___

57. Some movies: Abbr.

59. Old magazine ___ Digest

60. Fill in the blank with this word: ""Star ___""

62. Fill in the blank with this word: ""So ___ me!""

63. Worthless amount

64. Ka ___ (Hawaii's South Cape)

65. Some Mercedes-Benzes

PUZZLE 9

ACROSS

1. This single-celled fungus can ferment sugars & carbohydrates

6. Jon of TV's 'Homicide'

10. Would-be J.D.'s hurdle

14. United States biochemist (born in Spain) who studied the biological synthesis of nucleic acids (born in 1905)

15. Vladimir Nabokov novel

16. Fill in the blank with this word: "___ Aarnio, innovative furniture designer"

17. Some parade performers

19. Will who played Grandpa Walton

20. City rebuilt by Darius I

21. Kafka character Gregor ___

22. Fill in the blank with this word: "Bust ___"

23. Of a tart fruit: Prefix

25. Shoot over

27. Phony

30. Parent, e.g.

32. With 17-Down, a temporary urban home

33. Fill in the blank with this word: ""What ___ thou?""

35. Unloads

38. The English translation for the french word: Hel

39. Something left of center?

40. Park activity

42. Robert Louis Stevenson's "___ Triplex"

43. Unreadable

45. Folk/country singer Griffith

47. Scottish refusal

48. Some collar attachments

50. Sitting at a red light, say

52. Fill in the blank with this word: "___ Treaty, establishing the 49th parallel as a U.S. border"

54. Not so crazy

56. Gas: Prefix

57. Finish this popular saying: "You can have too much of a good_____."

59. Obama adviser Emanuel

63. Union member

64. On the rolls

66. Fill in the blank with this word: "___, zwei, drei"

67. Vases

68. Fill in the blank with this word: "___ fatuus"

69. Small paving stone

70. Jazzman ___ Allison

71. Stick out like ___ thumb

DOWN

1. Hebrew letters

2. Very light brown

3. Large food tunas

4. Comfort giver

5. Youngest world chess champion before Kasparov

6. The English translation for the french word: lance

7. Makes spoony

8. Turns down, in a way

9. Michael ___, Cochise player in 1950s TV

10. Fill in the blank with this word: "___ Lawrence Orchestra (British big band since the 1960s)"

11. Go out with the star of "The Wizard of Oz"?

12. March on, march on, since we ___ in arms': Richard III

13. This rich 5-letter cake with eggs, ground nuts & little to no flour, is down by contact with my stomach

18. Woman's shoe

24. Start of a quip

26. 1990s White House chief of staff Bowles

27. Sean Connery, for one

28. Uh-Oh! ___ (Nabisco product)

29. Queen's subject?

31. People: Prefix

34. Soprano ___ Huang

36. Sharon of 'Dreamgirls'

37. Go on a vacation tour

41. Hockey no-nos

44. Most provocative

46. Hands out, as duties

49. Anatomical cavity

51. Discount store offerings, for short

52. Where to find dates?

53. Tighten, as laces

55. This member of the parsley family has a distinctive licorice flavor & is an ingredient in ouzo

58. Guitar ___ (hit video game series)

60. The Ponte Vecchio crosses it

61. Fill in the blank with this word: "___ to the throne"

62. Warehouse contents: Abbr.

65. Fill in the blank with this word: "___ Maria"

PUZZLE 10

ACROSS

1. Some irregular sheets

11. Furman's partner in brokerage

15. Exobiologist's query

16. Wild Indonesian bovine

17. a native or resident of Tennessee

18. Parris Isl. outfit

19. Fill in the blank with this word: ""Take ___ a sign""

20. Uncle ___

21. Take ___ at (try)

23. Winery sight

24. Maxwell House alternative

26. Relating to tissue

28. Minneapolis suburb

30. Saccharin discoverer ___ Remsen

32. Fill in the blank with this word: "___ Onassis, Jackie Kennedy's #2"

33. Fine and dandy, in old slang

35. The ___ Bible

36. Potus #34

37. Water pipe in 16th-century Europe?

42. Fill in the blank with this word: "Bad ___, Mich. (seat of Huron County)"

43. The English translation for the french word: pleuvoir ‡ verse

44. The English translation for the french word: PIB

45. Fill in the blank with this word: "___ Jones"

46. Verb ending

47. Tasted, biblically

51. Skater Brian

53. The first one opened in Detroit in 1962

57. Veracruz Mrs.

58. XIII times XXXI

60. Without further ___

61. essential oil or perfume obtained from flowers

62. Tax ___

63. Needed things

66. Bone: Prefix

67. Listed

68. Library section

69. Roughly

DOWN

1. Fill in the blank with this word: "Cannabis ___ (marijuana)"

2. Icarus, e.g

3. Opera character who sings "Eri tu"

4. Barley beards

5. The Carolinas' ___ Dee River

6. Pari ___ (fairly)

7. 1942 Preakness winner

8. Spanish verse

9. Spanish queen until 1931

10. Sofer of soaps

11. Start to prepare, as 49-Across

12. It's tossed at Spanish restaurants

13. Winning coach of the first two Super Bowls

14. Fill in the blank with this word: "Country music's ___ Brown Band"

22. Visited overnight

24. Japanese chicken dish

25. Whim

27. Nothing, in Nice

29. R.V. stopover

31. Cockney greeting

34. Go-aheads

35. Fill in the blank with this word: "___ and cheese"

37. Swiss watch brand

38. someone who practices exorcism

39. Bulletin-creating department

40. Hide-and-___

41. File on an iPod

48. What's left behind

49. Takes to the soapbox

50. Units of capacitance

52. German indefinite article

54. Maestro Kurt ___

55. Stops on ___

56. Where to get a fast buck?

59. Fill in the blank with this word: "Alter ___ (exact duplicate)"

61. Uzbekistan's ___ Sea

62. Rock's ___ Lonely Boys

64. Rock's Brian ___

65. To ___ is human ...'

PUZZLE 11

ACROSS

1. African capital

6. '___ Breaky Heart"

10. Prefix for scope

14. Did a smithy's job

15. Mass in Arctic waters

16. Jack-in-the-pulpit, e.g.

17. Pleasant greeting

19. Voyeur's look-see

20. Bitty bark

21. Chowderhead

22. Flowering month

24. Media magnate Murdoch

26. Onslaught

30. Cyrano's prominent feature

31. Unstructured consciousness

32. Word with animal or a punch

34. "And let us not be ___ in well doing" (Gal. 6:9)

35. Ole's kin

36. Word with belly or ear

37. Fine partner

38. Glutton's request

39. Stubbed item

40. Spacek of the screen

41. Pre-1991 superpower

42. Nettle

44. Divine circle

46. Before the footlights

47. Stereotypes

49. Classification system for blood

50. Window base

51. Prickly seed vessel (Var.)

53. It'll hold water

56. Irritation or annoyance

59. Fruit with a wrinkled rind

60. Asian housemaid

61. River past Amiens

62. Powder substance

63. Winter Olympics event

64. Son of Japheth

DOWN

1. Covered with soot, e.g.

2. Mormon Tabernacle, for one

3. Horseback

4. 'Curse you, ___ Baron!"

5. Sun-dried bricks

6. Currently in progress

7. Layered sandwich

8. Masonry trough

9. Freeholders

10. Papyrus plant

11. Poet's "before"

12. Feel sorry about

13. Sitter's handful

18. Knight time?

23. Aesthetically pretentious

25. Jab playfully

26. Like some eyes

27. LBO for an exterminator?

28. Growls

29. Anesthetic of yore

31. Common thing

32. Place for a grilling

33. Chipmunk snack

34. Use inefficiently, as time

37. Slash

38. Backless slipper

40. Cutting thrust

43. Slanted, as type

44. Reception site, perhaps

45. Superlatively capable

47. Gracefully limber

48. Cum laude start

50. Dateless

52. Film spool

53. Word with up or down

54. Turkish title

55. Well, just the opposite?

57. Aussie avian

58. Old Cannes coin

PUZZLE 12

ACROSS

1. "___ bitten, twice shy"

5. Chafes

9. Cookbook abbr.

13. Oil source

14. Mideast hot spot

15. King or queen

16. Bride's personal outfit

18. Chilled

19. Bank offering, for short

20. ___ green

21. Back up

23. Starbucks selections

25. Period of the first dinosaurs

27. "Shoo!"

28. Experience

29. Bird ___

30. TV, radio, etc.

33. Mint, e.g.

36. races

38. Certain fir

40. Distort

41. www.yahoo.com, e.g.

42. "Comme ci, comme Áa"

44. "The Open Window" writer

48. Lustrous fabric

51. Movie preview

53. Ambiguous or unclear

54. Hack

55. Caribbean, e.g.

56. Dine at home

57. Stranded at Sugarloaf

60. Certain sorority member

61. Beach bird

62. At liberty

63. New England catches

64. Go to and fro

65. "Bill & ___ Excellent Adventure"

DOWN

1. Ideals

2. Thin

3. The cavity

4. Australian runner

5. Gets promoted

6. Component used in making plastics and fertilizer

7. Bleat

8. Herald, for one

9. One of the Barbary States

10. In seventh heaven

11. Capable of being cut

12. Grand ___ ("Evangeline" setting)

15. Santa ___, Calif.

17. Marienbad, for one

22. Flint is a form of it

24. Sacred songs

25. Mob disperser

26. Blackguard

28. Alpine sight

31. In-flight info, for short

32. Bust

34. "Dear" one

35. Chip dip

36. Doomed

37. "___ calls?"

38. Except

39. Fox relative

43. A limestone

45. Cinch

46. Lamented

47. A decree of a Muslim ruler

49. Moves erratically, as a butterfly

50. "Empedocles on ___" (Matthew Arnold poem)

51. Lion-colored

52. "Chicago" lyricist

54. "The Last of the Mohicans" girl

56. "Yadda, yadda, yadda"

58. "What's ___?"

59. "How ___ Has the Banshee Cried" (Thomas Moore poem)

PUZZLE 13

ACROSS

1. Cig

6. Bad day for Caesar

10. The Sail (southern constellation)

14. Leg bone

15. Astronaut's insignia

16. Persia, now

17. About

18. Assign great social importance to

20. Mayor of a municipality in Germany

22. Baby's first word, maybe

23. "Seinfeld" uncle

24. "What's ___?"

25. "... ___ he drove out of sight"

26. Addition

27. "Empedocles on ___" (Matthew Arnold poem)

29. Boris Godunov, for one

31. "Mi chiamano Mimi," e.g.

32. In favor of

34. An organization of missionaries in a foreign land

37. Operating on living animals

39. Crusader's foe

40. www.yahoo.com, e.g.

41. ___ function

42. Jewish month

44. Charged particles

48. Bank offering, for short

49. Amigo

51. "Silent Spring" subject

53. Brouhaha

54. Antiquity, in antiquity

55. Containing or resembling amethyst

58. Action takeing place during a road journey

60. Up, in a way

61. Baptism, for one

62. Bad look

63. Cooktop

64. Again

65. "... or ___!"

66. Positions

DOWN

1. It'll hold your horses

2. Ballroom dance

3. Moon of Uranus

4. Double-decker checker

5. Big name in stationery

6. Marriage acquisition

7. Honoree's spot

8. Abstruse

9. More rational

10. Relative of "i.e."

11. Cosmopolitan genus of usually perennial herbs

12. Ointment ingredient

13. Neighbor of Namibia

19. Anger

21. Kind of unit

28. All excited

30. Change, as the Constitution

31. Garlicky mayonnaise

33. Egg cells

35. Well-built

36. "Dear" one

37. A salt or ester of vanadic acid

38. It rises every year

39. THEME ANSWER 5

41. Code word for "S"

43. Cleave

45. Familiarize

46. Minority

47. Stockholm natives

49. "Polythene ___" (Beatles song)

50. Plant used as soap

52. Autocrats

56. Associations

57. Asian tongue

59. Drops on blades

PUZZLE 14

ACROSS

1. Falling flakes
5. Commoner
9. Bridal path
14. Novice
15. "Where the heart is"
16. Fat
17. Biblical birthright seller
18. "Not on ___!" ("No way!")
19. Allotment
20. Means of support
22. Big Indian
24. Dusk, to Donne
25. In a dim indistinct manner
27. Decorated, as a cake
29. Watergate, e.g.
32. Bulbous herb of southern Europe
35. Egyptian Christian
36. Sees
39. Hokkaido native
40. Increase, with "up"
41. 1935 Triple Crown winner
42. Fold, spindle or mutilate
43. Page
45. Small tree native to the eastern United States
46. Durable wood
47. A sculpture representing a human or animal
49. Wore
51. ...
52. Purple shade
53. "Bingo!"
55. ___ Strip
57. Rubenesque
61. Black
63. Court attention-getter
65. ___-bodied
66. ___ wrench
67. Elliptical
68. Poker action
69. Fold
70. ___ mortals
71. At one time, at one time

DOWN

1. Increase, with "up"
2. Not yet final, at law
3. Face-to-face exam
4. Would not
5. a book containing a compilation of pharmaceutical
6. Court ploy
7. File
8. ___ carotene
9. Petting zoo animal
10. "Rocky ___"
11. Top speed
12. Vermin
13. "Our Time in ___" (10,000 Maniacs album)
21. A hand
23. A puzzle
26. Andean animal
28. Victorian, for one
29. Picket line crossers
30. Cleanser brand
31. Capable of being pacified
33. Absurd
34. Fixed
37. Drink from a dish
38. ___ apso (dog)
44. "Harper Valley ___"
46. Introduce
48. Pressing
50. Egg cells
52. May have
53. "By yesterday!"
54. Campus building
56. PBS show "by kids, for kids"
58. Alpine transport
59. Misfortunes
60. Cabbage
62. Grassland
64. Ring bearer, maybe

PUZZLE 15

ACROSS

1. Priestly garb

5. ...

10. Speech problem

14. ...

15. Minimal

16. A chip, maybe

17. "American ___"

18. Job holder?

19. Boris Godunov, for one

20. ...

22. Relating to or extending over

24. "The dog ate my homework," e.g.

27. ¿ la mode

28. Bit of paronomasia

30. "Giovanna d'___" (Verdi opera)

31. Bully

34. ___ Wednesday

35. Like Beethoven

36. Like "The X-Files"

37. Highlander

39. Like a rainbow

42. "To thine own ___ be true"

43. "Encore!"

45. "Your turn"

47. ___ Dee River

48. Difficult or unpleasant

50. "Scream" star Campbell

51. "To ___ is human ..."

52. Binge

53. Go places

55. Crack

58. Small evergreen shrub of Pacific coast of North America

61. Calf-length skirt

62. Force units

65. Sacred Hindu writings

66. "Cast Away" setting

67. Affect

68. Aims

69. Quite a while

70. English exam finale, often

71. Eye affliction

DOWN

1. "___ I care!"

2. Disney dog

3. In a broken-hearted manner

4. A pure form of finely ground silica

5. Priestly garb

6. "Fantasy Island" prop

7. Reading room furniture?

8. Christiania, now

9. Fetor

10. ...

11. A chemical substance that repels insects

12. Antares, for one

13. Make waves

21. Bunch

23. Form of clarified butter

25. Component used in making plastics and fertilizer

26. Cicatrix

28. Heathen

29. Grammar topic

32. Edmonton hockey player

33. Allude

38. Menservants and chauffeurs

40. "... happily ___ after"

41. Vedic mythology

44. "Cool!"

46. Guns

49. One who works hard at boring tasks

54. Overhangs

55. During

56. Galileo's birthplace

57. Bakery selections

59. Email contact info, in slang

60. Beam intensely

63. In-flight info, for short

64. ___ sauce

PUZZLE 16

72. Way: Abbr.

73. Feeling, Italian-style

DOWN

1. Old draft deferment category for critical civilian work

2. Hightailed it

3. Who's there?' answer

4. Woman's name suffix

5. Spend

6. Sell door-to-door

7. "The one-l lama," to Ogden Nash

8. Actress Madlyn

9. 1957 Literature Nobelist

10. San ___, Argentina

11. Rare blood designation

12. Sammy Davis Jr. had one

13. Fill in the blank with this word: ""The ___ trick in the book!""

21. Combat with fighter-bombers

25. Starts to raise, as a hem

26. Teacher's deg.

27. Fill in the blank with this word: "___-la-la"

28. Rembrandt van ___

29. Time of legend

31. Word repeated in Emily Dickinson's "___ so much joy! ___ so much joy!"

34. The English

translation for the french word: tremble

36. Mark your card!

38. Wooed very well

40. Range part: Abbr.

41. Fill in the blank with this word: ""___ was saying Ö""

42. Fill in the blank with this word: "___-tac-toe"

43. In tune

45. Panoramic photos

47. Record keeper

48. Suitcase convenience

49. Mark who won two majors in 1998

50. Most widespread

52. Start of a play?

55. Apple tablet computers

57. Flood barriers: Var.

60. Unite formally

61. Cyclops' feature

62. The Last Time I Saw Paris' composer

63. Fill in the blank with this word: ""Breaking ___ Hard to Do""

64. San ___, Calif.

ACROSS

1. Vietnam War name

6. Fill in the blank with this word: "___ zoologique (French zoo)"

10. Just let ___'

14. Olympian Katarina et al.

15. Biblical dry measure: Var.

16. Word div.

17. Spanish direction

18. Tom-tom

19. ___ Pro (2015 debut)

20. Words of farewell from Childe Harold

22. Not those, in Brooklyn

23. Work unit

24. Uses a hose

26. Sein : German :: ___ : French

30. Finish this popular saying: "A change is as good as a_____."

32. What 'check' could mean

33. Prohibition of alcohol sales

35. Cookie containers

37. Shop tool with pulleys

39. Tone-blending painting technique

44. Whizzes

46. Doesn't decline

47. 10,000,000 rupees, in India

51. Va. neighbor

53. Small cut

54. Mater ___ (Mary, in Latin prayers)

56. Town line sign abbr.

58. Lee Van ___ (spaghetti western actor)

59. "Li'l Abner" mother

65. Zoological wings

66. Fill in the blank with this word: "Et ___"

67. Fill in the blank with this word: "___ low profile"

68. Sun. talks

69. Deceptively manipulate, with "up"

70. Isolated nest: Var.

71. Quod ___ faciendum

PUZZLE 17

ACROSS

1. Math amts.

5. Teacher's advanced deg.

9. Fill in the blank with this word: ""Cûmo ___?""

14. Fill in the blank with this word: "Elaine ___, George W. Bush's only labor secretary"

15. Wildcats' org.

16. Singer Aguilera, self-referentially

17. It gets stuck in bars

19. Antigone's cruel uncle

20. Avis: "We ___ harder"

21. Wont

22. Theseus abandoned her

24. Widespread damages

26. Fill in the blank with this word: "___ minÈrale"

27. Yellowish pink color

31. Things associated with pits and spits, briefly

35. Hindu sage

38. Biblical spy

40. Moli√°re's 'Le M√©decin Malgr√© ___'

41. Cooking

44. Fill in the blank with this word: "Dynamic ___"

45. Fill in the blank with this word: "___ newt (witch's ingredient)"

46. Thomas Mann's "___ Kroger"

47. Workers need them: Abbr.

49. Hands out, as duties

51. Whole lot

53. Two of the three gifts of the Magi

57. One reason to get a cross

61. Popular 1980s arcade game based on simple geometry

62. Emergency call

63. Fill in the blank with this word: ""I Didn't Slip, I Wasn't Pushed, ___" (1950 song)"

64. *Mark the transition from an old year to the new, maybe

67. Stick fast (in)

68. This fermented honey-&-water beverage was a favorite of Chaucer's miller & of the god Thor

69. Hit a ___

70. Threesome

71. They're unaffiliated: Abbr.

72. Simon who wrote "The Death of Napoleon"

DOWN

1. Entertainment center at many a sports bar

2. Office building cleaner

3. Pot

4. Wobbly walker

5. Alphabet sextet

6. Word to a tabby

7. Tin ___

8. Word from the crib

9. No longer working for the Company

10. Horror novelist Peter

11. Undecided, you might say

12. I'll come to you ___': Macbeth

13. Within reason

18. Certain reed

23. Spot remover?

25. Saudi monarch

28. They're served with spoon-straws

29. Underworld bosses

30. Viking king, 995-1000

32. Buckwheat pancake

33. Je ne sais ___

34. Fill in the blank with this word: "___-Tibetan languages"

35. Some mail designations: Abbr.

36. Written promises

37. Fill in the blank with this word: ""So ___?""

39. Be on deck

42. Old soap '___ Hope'

43. Unscramble this word: stso

48. Her feast day is Jul. 11

50. Slavish routines

52. Tipsy

54. Hole-___

55. "Forget it!"

56. Numbers

57. Sword handle

58. Give an ___ effort

59. Fill in the blank with this word: "___-Tea (first instant iced tea)"

60. Year Queen Victoria died

61. View from the dorms

65. Wine: Prefix

66. WWW access option

PUZZLE 18

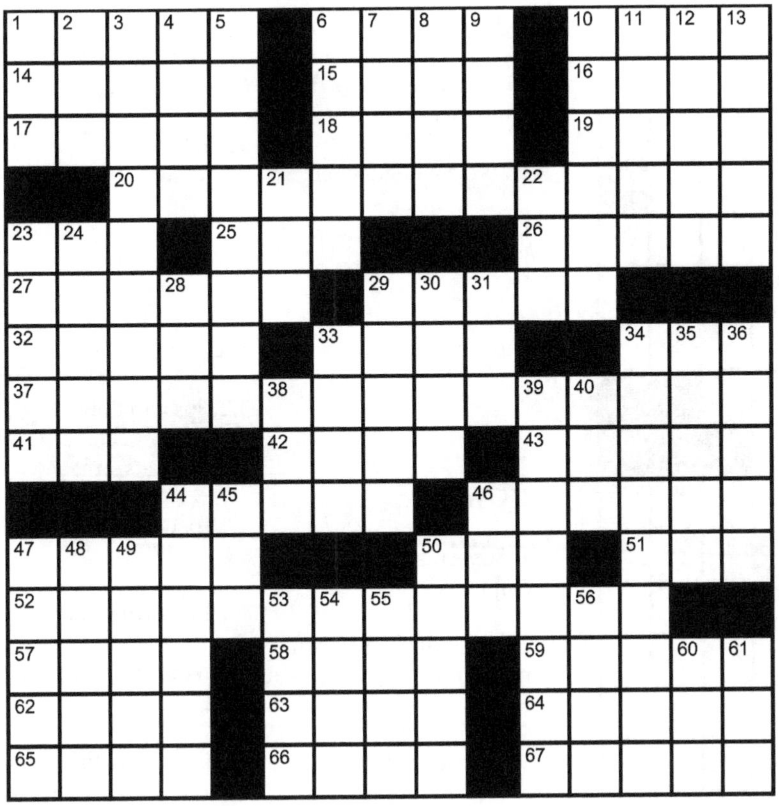

The crossword grid with numbered cells.

ACROSS

1. Train stop

6. Fill in the blank with this word: "Fashion's ___ Saint Laurent"

10. Peek-___

14. Declarer

15. The English translation for the french word: groin

16. Fall mos.

17. Fill in the blank with this word: ""That ___ lady ..."

18. On bended ___

19. Fill in the blank with this word: ""___ Death" (Grieg work)"

20. It's classified

23. Sorority letters

25. Withheld

26. Fill in the blank with this word: ""___ sorry!""

27. Stuck

29. Key of Schubert's Symphony No. 5

32. Visit

33. South American monkey

34. Some undergrad degs

37. Features found in 17- and 64-Across and 11- and 28-Down

41. Fill in the blank with this word: "___ out a win"

42. Explorer John and others

43. Play co-authored by Mark Twain

44. Made off with

46. Provincial capital in NW Spain

47. Year's record

50. Fill in the blank with this word: "___ sponte (of its own accord, at law)"

51. Tight end, at times

52. Concerns of Archimedes

57. See 70-Across

58. Word-of-mouth

59. Parts of masks

62. Potato source

63. Mother and wife of Uranus: Var.

64. Hawaiian feasts

65. Ward on TV

66. Three-stripers: Abbr.

67. Raptor 350 and others

DOWN

1. Thomas Bailey Aldrich story "Marjorie ___"

2. Rice and Lloyd Webber's "Waltz for ___ and Che"

3. Quickly

4. Wine: Prefix

5. Den decorations

6. On the team?

7. Price of a movie?

8. D.O.E. part: Abbr.

9. Fill in the blank with this word: "___-Ball"

10. Fill in the blank with this word: ""Vigilant ___ to steal cream": Falstaff"

11. Twiggy broom

12. Subjective pieces

13. Bone: Prefix

21. Fill in the blank with this word: "___-string"

22. Fill in the blank with this word: "___ Maria"

23. Watchmaker's unit of thickness

24. Thumbing-the-nose gesture

28. Late ___

29. Snake, for one

30. Quarters

31. Finish this popular saying: "Let sleeping dogs ___."

33. Purveyor of nonstick cookware

34. Nine

35. Fill in the blank with this word: ""___ always say ..."

36. What to call un hombre

38. Tulsa sch. named for an evangelist

39. How bitter enemies attack

40. Fill in the blank with this word: ""The Sweetheart of Sigma ___"

44. Violinist Heifetz

45. She, in S

46. Arles assent

47. Sap-sucker's genus

48. [See title, and proceed]

49. Mass communication?

50. Ward and namesakes

53. Yuletide quaffs

54. Zip strip?

55. Fill in the blank with this word: ""Hurry up and ___""

56. Sign of a hit show

60. Mao's successor as Chinese Communist leader

61. Worrying sound to a balloonist

PUZZLE 19

ACROSS

1. Afterthought #3: Abbr.

5. Crackerjack

10. Ming's 7'6" and Bryant's 6'6", e.g.: Abbr.

14. Explorer John and others

15. ___ deck

16. Rude audience member

17. Among those attending

19. Town line sign abbr.

20. Fill in the blank with this word: "Elevator ___"

21. Veg out

22. Pancreatic enzyme

24. Suffix with morph-

25. War on Poverty agcy.

26. Very much

28. With 6- and 22-Across, noted 19th-century writer

30. Not be able to swallow

32. Word with boss or bull

33. R & B group with the 1991 #1 hit "I Like the Way"

35. TV announcer Hall

36. "Ugh!"

37. Pep, doubled

40. Arg. neighbor

42. Italian ___

43. The 21st, e.g.: Abbr.

44. Weather London is famous for

45. Jawbone of ___ (biblical weapon)

47. Words of explanation

51. Genus of poisonous mushrooms

53. W. C. Fields film "___ a Gift"

55. Fill in the blank with this word: "Anderson's "High ___"

56. Fill in the blank with this word: ""All systems ___"

57. Snick-or-___

58. '___ you to horse': Macbeth

59. Combined, in Compi

60. Addleheads

63. Fill in the blank

with this word: "___ Piper"

64. Eye: Prefix

65. Used bookstore containers

66. Voyager launcher

67. Maker of rifles and revolvers

68. Fill in the blank with this word: "___-deucy"

DOWN

1. Supermarket employee

2. Fill in the blank with this word: "___ City, Fla."

3. Tube-nosed seabird

4. Secure online protocol

5. Tough row ___

6. The English translation for the french word: prologue

7. Month preceding Rosh Hashanah

8. Whoop-de-___ (big parties)

9. The English translation for the french word: apÈriodique

10. Physical sound?

11. Dirt spreader

12. Distinctly representative

13. Non sibi ___ patriae' (Navy motto)

18. Old Testament God

23. Stitch

26. Understands

27. Ordinal suffix

29. Illustrator for Charles Dickens

31. Yellow-fever mosquito

34. What a sensitive nose may detect

36. Over there, poetically

37. Ups and downs

38. Wildcats' org.

39. Theodor ___ (Dr. Seuss)

40. Suffix with odd

41. Where the Danube ends

45. Since 2008 this insurance conglomerate has gotten $182 billion in government funds

46. TV's "Living ___"

48. Unscramble this word: nihcet

49. Elegantly groomed

50. Like Rapunzel

52. I ___ vacation!'

54. Fill in the blank with this word: ""Coffee, ___ Me?""

57. Tightfitting

59. WB competitor

61. Smallest NATO member by population

62. Hall-of-Fame basketball coach Hank

PUZZLE 20

ACROSS

1. Fill in the blank with this word: ""___ la vie""

5. The Bourne Identity' director Liman

9. Pointing a finger at

14. Small and insignificant

15. Venezuela's ___ Margarita

16. Like some cigarettes

17. Quite limber

20. Some clouds

21. One-inch pencil, say

22. Some Bourbons, par exemple

23. Isn't ___ bit like you and me?' (Beatles lyric)

24. Years, to Yves

26. Fill in the blank with this word: "___ and proper"

28. Unspecified degrees

30. Boss's words, after 'Because'

34. Give nothing to

37. Hollywood's Roberts and others

39. Wolfed down

40. 1908 Cubs player and position

44. Fill in the blank with this word: ""___ Wanna Cry," 1991 #1 song"

45. Amazes

46. Vol. 1, No. 1, e.g

47. Actresses Shire and Balsam

49. Vapory beginning

51. Title role for Chris Hemsworth

53. Torah place marker

54. Writer ___ Louise Huxtable

57. Research facility: Abbr.

60. The English translation for the french word: DIU

62. Cool, man!'

64. Bodies making their closest approach in more than 50,000 years on August 27, 2003

67. Amazon parrot

68. Inventor Elias

69. '60s protest / Skip, as a dance

70. Capital city about an hour by plane from Miami

71. Site of Ikea's hdqrs.

72. Variety of agate

DOWN

1. Team leader

2. Succeed

3. 1965 march site

4. The personal pronoun of the second person in the nominative case; people often act holier than it

5. Raise the dead?

6. Verb ending

7. Fill in the blank with this word: "___-Ude (Trans-Siberian Railroad city)"

8. Refuel

9. Paderewski's "Minuet ___"

10. Easily-used people

11. Was ___ hard on them?'

12. Nose: Prefix

13. Tests for srs.

18. Mechanic's ___

19. Dugout shelter

25. Fill in the blank with this word: "___ kebab"

27. Fill in the blank with this word: ""Time ___ a premium""

29. Vowel sound

31. The English translation for the french word: yÈti

32. Tomato and vegetable

33. Pos. and neg., e.g.

34. "Cease and desist!"

35. Ricky Martin's "Livin' La ___ Loca"

36. Keto-___ tautomerism (organic chemistry topic)

38. Sentimentalist

41. Jersey workers

42. Greenland base for many polar expeditions

43. Tarried

48. Fill in the blank with this word: ""Bon ___""

50. Weirdo

52. Fill in the blank with this word: "___ Chris Steak House"

54. Fill in the blank with this word: ""Not ___!""

55. X-rated

56. When some watch the local news

57. Fill in the blank with this word: "Call ___ evening"

58. Dragsters' org.

59. Wet septet

61. Red Sea vessel

63. The Pointer Sisters' "___ Excited"

65. Fill in the blank with this word: ""___ approved" (motel sign)"

66. In ___ of

PUZZLE 21

ACROSS

1. Distinctive style

6. Turns on the waterworks

10. Word to a fly

14. Gold unit

15. Healthy berry

16. Condor's pad (Var.)

17. Surface layer

19. Unfettered

20. Archer's wood

21. To boot

22. Beat a dead horse

24. Farm baby

25. Kuwaiti prince

26. Piece†of†work

31. Doohickey

32. Chaucerian tale-teller

33. 'Ay, there's the ___"

35. More green around the gills

36. 'Dead man's hand" card

37. Coward's ''To Step ___"

39. 'My ___" (Mary Wells classic)

40. Jamaican pop music

41. Wince

42. likeness or counterpart

46. Moonshine ingredient

47. Bit of nuclear physics

48. He gets what's coming to him

51. One-customer connector

52. Baseball great Mel

55. Something fishy?

56. Evergreen perennial

59. Gal's sweetheart

60. Personal appearance

61. Celery

62. Wraps up

63. Ratted

64. Disintegrates, as a cell

DOWN

1. Fuddy-duddy

2. Inveigle

3. Yet again

4. Cooler cooler

5. A rented car

6. Prune

7. 'L'___ del Cairo" (Mozart opera)

8. Grill or fire

9. Jazz band instrumentalist

10. Hunting journey, often

11. Basil or sage, e.g.

12. Two-colored cookie

13. ___ and terminer

18. Like a rock

23. Torched

24. Relinquish control

26. Alone

27. Tentacled creature

28. Machu Picchu denizen

29. NASA problem part

30. Urge physically, but gently

31. Archaeological excavation

34. One who's always buzzing off?

36. Motionlessness

37. Heraldry or heraldic

38. Famous twins' birthplace

40. Mountain waterways

41. Quoted as an authority

43. His wings melted in the sun

44. Little tyke

45. Wonder or awe

48. Solitaire spot, perhaps

49. Famous apple site

50. Prompt

52. Singer Redding

53. Undeniable

54. Highland hats

57. It's mightier than the sword

58. Finder of secrets

PUZZLE 22

61. On the safe side, at sea

62. History Muse

63. Anatomical dividers

64. Frau's partner

65. Suspended

DOWN

1. Arrogant and annoying

2. Look angry or sullen

3. The "A" of ABM

4. A constellation in the southern hemisphere

5. Charity event in the park

6. Provide, as with a quality

7. Gulf of ___, off the coast of Yemen

8. Anderson's "High ___"

9. Acquired relative

10. For the most part

11. Fasten

12. Component used in making plastics and fertilizer

13. All ___

18. "Jo's Boys" author

21. Cell alternative

23. "Carmen" composer

24. Play, in a way

25. Bouquet

26. Eccentric

27. Ornamental flower, for short

28. Small woods

30. Avoid

31. Composer Copland

32. Howler

34. Alpha's opposite

37. Clear, as a disk

38. Dismays

39. ___ el Amarna, Egypt

44. During

46. Elastic

48. Swelling

49. More despicable

50. Express Mail org.

51. Dermatologist's concern

52. Barber's motion

53. Stubborn beast

54. Allergic reaction

55. Yellowish brown balsam

56. "___ Brockovich"

57. E.P.A. concern

59. Code word

ACROSS

1. Bohemian, e.g.

5. Accomplishment

9. Permeate

14. ___ of the above

15. Annul

16. ...

17. From a foreign country

19. Lid or lip application

20. See if or how it works

21. National Zoo favorites

22. Nod, maybe

23. Reprimand, with "out"

24. Yellowish pink

28. Butt

29. "Awright!"

33. Rainbow ___

34. Exude

35. Ballad

36. Normal temperature of room

40. Person in a mask

41. Medical advice, often

42. Deceived

43. Poet Angelou

45. "Act your ___!"

46. Lead source

47. ___ Verde National Park

49. Keep out

50. Advantages

53. Soup cooked in a large pot

58. Balloon probe

59. A windstorm

60. Geometrical solid

PUZZLE 23

ACROSS

1. CaffÉ ___

6. Apprehension

11. Fed. construction overseer

14. ___ squash

15. Like "The X-Files"

16. Amscrayed

17. An English verb

19. "A jealous mistress": Emerson

20. Hammer part

21. Born, in bios

22. Indian coin

24. Ten-armed oval-body

28. Biker's hot-dog maneuver

31. Length x width, for a rectangle

32. Verb with thou

33. A rude decoration

37. ___ be an honor'

38. Kind of room

40. Calendar abbr.

41. Adjust anew

44. Barely speak

46. Christmas season

47. By and large

49. A preview to test audience reactions

53. Crush

54. Cashew, e.g.

55. Shot in the arm

59. Ashes holder

60. Fight with Oscar winner Sally?

64. ì___ boom bah!î

65. "La BohËme," e.g.

66. Barely beat

67. Pillbox, e.g.

68. Roentgen's discovery

69. CÈzanne contemporary

DOWN

1. Arctic native

2. Advil target

3. Bolted

4. A psychological state

5. "Star Trek" rank: Abbr.

6. Computer key

7. Celebrate

8. "... ___ he drove out of sight"

9. Balloon filler

10. Debriefing and make him report

11. Result of some plotting

12. Eastern wrap

13. Chips in

18. The "A" of ABM

23. Org. that uses the slogan 'Aim High'

25. Final: Abbr.

26. All ___

27. Phi Delt, e.g.

28. Blender sound

29. "Unimaginable as ___ in Heav'n": Milton

30. Icelandic epic

33. "___ lost!"

34. "American ___"

35. Blue hue

36. Gross

38. Mouthful

39. ___-friendly

42. Pair

43. Singles player

44. Nod, maybe

45. Series on which Clint Eastwood played Rowdy Yates

47. Throat hangers

48. Locale

49. "Pipe down!"

50. Water wheel

51. Artist Max

52. Admittance

56. Burglar

57. "Guilty," e.g.

58. "What are the ___?"

61. 30-day mo.

62. Oolong, for one

63. Disobeyed a zoo sign?

PUZZLE 24

ACROSS

1. Sea eagles

5. Increases greatly, as prices

9. Welcome sight for a castaway

14. Fill in the blank with this word: "___ de soie (silk cloth)"

15. Fill in the blank with this word: ""Die Frau ___ Schatten" (Strauss opera)"

16. Slightly above average

17. Kind of sale

19. Trifles: Fr.

20. White ___ ghost

21. 1977 double-platinum Steely Dan album

22. Picks up

23. Mandela's land: Abbr.

24. Mystery writer Gardner et al.

26. Swell

29. Redundancies, like 20- and 50-Across and 5- and 29-Down

33. Have ___ for (desire)

34. Fatty

35. Withdraw gradually

36. Food flavoring brand

37. Wielder of the sword Tizona

38. Vapory beginning

39. Fill in the blank with this word: "___-Honey (candy name)"

40. Opera singer Simon ___

41. "That was close!"

42. Water coolers

44. The English translation for the french word: piste de ski

45. Fill in the blank with this word: ""___ fast!""

46. Union ___: Abbr.

47. Gems, precious metals, etc., in Spain

50. Writer Santha Rama ___

51. School media depts.

54. The English translation for the french word: trait

55. See 20-Across

58. Ukrainian port, to natives

59. Answer to the riddle "Dressed in summer, naked in winter"

60. Tennis's Mandlikova

61. Lulls

62. Supplementary: Abbr.

63. She-bears, south of the border

DOWN

1. Biblical dry measure: Var.

2. Actor Stephen et al.

3. "The Lion King" lion

4. "-er" or "-ing," e.g.: Abbr.

5. Sara portrayer on 'CSI'

6. Some Muslims

7. The Tar Heels: Abbr.

8. The Carolinas' ___ Dee River

9. Window insert

10. Actress Zadora visited Samoa's capital?

11. The Sopranos' actor Robert

12. Sally ___ (teacake)

13. Sound of a leak

18. Yesteryear

22. London insurance giant

23. EIOSN

24. Villain of Spider-Man

25. Knots again

26. Fill in the blank with this word: ""___ Meets Godzilla" (classic 1969 cartoon)"

27. With 39-Across, 21-/28-Across, for one

28. Spanish direction

29. Beats

30. Actor Green and others

31. "The Old Religion" novelist

32. Senator from Maine

34. Fill in the blank with this word: ""Never ___ tell thy love": Blake"

43. Main lines

44. Spittoon sound

46. Savanna region stretching from Senegal to Chad

47. 'Vette option

48. Wagner's earth goddess

49. Writer's supply: Abbr.

50. Snow White's sister

51. Fill in the blank with this word: ""___, 'tis true I have gone here and there": Shak."

52. Fill in the blank with this word: "___ cava"

53. Women of Andaluc

55. This abbreviation .gov promises that the organization is "bringing safety to America's skies"

56. La Guardia : LGA :: O'Hare : ___

57. Notwithstanding that, informally

PUZZLE 25

ACROSS

1. Biblical ed.

4. Have a ___ stand on

9. Tissue: Prefix

14. Fill in the blank with this word: "Capitol-___ (music company)"

15. You should whip this ingredient before you top your Chantilly potatoes with it

16. Overindulgent parent, e.g.

17. Title for this puzzle

20. Use a knife

21. Sign for May Day babies

22. Gossipy group

26. Word modifier: Abbr.

27. Word with boss or bull

30. PBS's '___ Can Cook'

31. Mardi Gras, e.g.: Abbr.

33. Fill in the blank with this word: "___ Field"

35. Limestone regions with deep fissures and sinkholes

37. Fill in the blank with this word: "___ : hello :: hooroo : goodbye"

38. Feature of 20- and 35-Across, forward and backward

42. Singer Turner

43. 13th-century king of Denmark

44. Like many a grandparent

47. Wrangle

48. Line score letters

51. West Bank grp.

52. Literary monogram

54. Pussyfooted

56. in gear (similar term)

59. Ohio native

60. California city by Joshua Tree National Park

65. Tubular pasta

66. Infection fighter

67. Rock's ___ Lonely Boys

68. Leo with the 1977 #1 hit 'You Make Me Feel Like Dancing'

69. The English translation for the french word: Ègide

70. Warbler Sumac

DOWN

1. Subject of a sailor's weather maxim

2. Title character in a Peter Hoeg best seller

3. Vance of "I Love Lucy"

4. Rock's ___ Soundsystem

5. We'll teach you to drink deep ___ you depart': Hamlet

6. What a H.S. dropout may get

7. Fill in the blank with this word: ""___ teaches you when to be silent": Disraeli"

8. Mutual of ___

9. Stock ticker inventor

10. Whit

11. Western festival

12. Fill in the blank with this word: "___ el Amarna, Egypt"

13. Surgery sites, for short

18. TV's Cousin ___

19. Word ending meaning "foot"

23. Union member

24. Weightlifter's rep

25. Wiesbaden's state

28. Fill in the blank with this word: "___-Tass news agency"

29. Stowe girl

32. Wakens

34. Understands

35. Fill in the blank with this word: ""The Bridge on the River ___""

36. Sounds that may be heard before bangs?

38. Six-stringed instrument

39. a purely biological unfolding of events involved in an organism changing gradually from a simple to a more complex level

40. Garage container

41. To laugh, to Lafayette

42. Group formed at C.C.N.Y. in 1910

45. More elegant

46. With 29-Down, central role on "Knots Landing"

48. Singer with the 1994 #1 hit "Bump N' Grind"

49. Child's attention-getting call to a parent

50. One of the Crusader states

53. Tetra- plus one

55. Fill in the blank with this word: ""The ___ Daba Honeymoon""

57. Trix alternative?

58. Fill in the blank with this word: ""No ___!""

60. Pulls a certain prank on, informally

61. Fill in the blank with this word: ""Are ___ pair?" ("Send in the Clowns" lyric)"

62. Photog's item

63. Ransom ___ Olds

64. Some nouns: Abbr.

PUZZLE 26

ACROSS

1. Was artificially cooled, for short

6. Fill in the blank with this word: "___ lot (gorged oneself)"

10. Jazz singer ___ James

14. Eyes

15. Turkish brandy

16. India's ___ Jahan

17. Tony-, Oscar- and Emmy-winner born 10/10/1900

19. Peacock constellation

20. The appendix extends from it

21. Bit of a muscle car's muscle

23. What ___ told you ...?'

25. Terrifying tales

28. Fill in the blank with this word: "Also-___ (losers)"

30. Whittier war poem "Laus ___"

31. Greene of "Bonanza"

32. Faith in God

35. Stage that includes a cocoon

37. Ivy League rooters in green

42. Mil. unit below a division

43. Cobra products

45. French composer Erik

49. Fill in the blank with this word: ""Did you ___ that?""

51. Fill in the blank with this word: "___ Pet (novelty item)"

52. Line in London

56. You might take stock in it: Abbr.

57. Typos

58. Honey badger

60. Fill in the blank with this word: "Dragon's ___ (early video game)"

61. 2007 satirical best seller subtitled "And So Can You!"

66. "Fatal Attraction" director Adrian

67. Proper ___

68. What a tragedy!'

69. USMC rank

70. Side-channel, in Canada

71. Wreck

DOWN

1. Water, chemically

2. Whiz-bang

3. Don Quixote's do

4. You'll use up 3 vowels playing this word that means toward the side of a ship that's sheltered from the wind

5. Sure thing

6. Synthetic polyamide used in fiber-making

7. Scotland's Firth of ___

8. Fill in the blank with this word: "___ out a win"

9. Tyler of 'Ghost Whisperer'

10. Husband, in Hidalgo

11. De-icer

12. The English translation for the french word: taverne

13. What Richard III offered "my kingdom" for

18. Mao's successor as Chinese Communist leader

22. Seasoned rice dishes

23. Fill in the blank with this word: ""Just Another Girl on the ___" (1993 drama)"

24. Saudi monarch

26. Ostensible

27. Wipeout

29. Yes, ___'

33. Springsteen's E ___ Band

34. Woman of la maison: Abbr.

36. Chem. class measures

38. Today, in Turin

39. Westminster Show org.

40. someone who rejects all theories of morality or religious belief

41. Use tiny scissors

44. Fill in the blank with this word: "___ TomÈ"

45. Witches' recitations

46. Lines up neatly

47. Strenuous

48. Turkish inn

50. Capital whose central plaza is Skanderbeg Square

53. Torrents

54. Texas ___

55. Weapons check, in brief

59. Switch suffix

62. Get an ___ (ace)

63. Fill in the blank with this word: ""___ bien""

64. Mike Ovitz's former co.

65. The English translation for the french word: AVQ

PUZZLE 27

ACROSS

1. Natl. Boss Day, ___ 16

4. Terrif

7. World view

10. Satellite ___

13. It premiered the day before "E.R.": "___ Hope"

15. Muhammad ___

16. Fill in the blank with this word: "Electric ___"

17. Container for preserving historical records

19. Unruly crowd

20. Unrushed pedestrian

21. Exaggerate

23. The English translation for the french word: sinople

24. The English translation for the french word: fidÈlitÈ

28. Fill in the blank with this word: ""Able was I ___ Ö""

29. The English translation for the french word: moi

30. Observation

31. The English translation for the french word: imago

33. "Listen!"

34. Things to "see" in an encyclopedia

40. Fill in the blank with this word: "___ of Sandwich"

41. Threefold

42. Untrustworthy types

45. Fill in the blank with this word: "___ and outs"

46. Unscramble this word: opp

49. Any nonverbal action or gesture that encodes a message

52. Water carrier

53. Yo, this pope III had a "Rocky" one-year reign from 884 to 885

54. This job can also mean to fashion your actions to the needs of another

56. Zine reader

58. A public relations person

60. The English translation for the french word: frangin

61. Sound of contempt

62. The English translation for the french word: roucouyer

63. Fill in the blank with this word: ""___ true!""

64. The work of Kim Philby & "Harriet", or to catch sight of suddenly

65. Worker in a garden

66. Finish this popular saying: "Like father, like_____."

DOWN

1. The English translation for the french word: octave

2. Bellman

3. This quality of sound distinguishes it from other sounds of the same pitch & volume

4. Finish this popular saying: "Monday's child is fair of_____."

5. Petri dish gel

6. Thwack

7. The English translation for the french word: violet

8. The English translation for the french word: allÈguer

9. Type genus of the Pieridae

10. One who may get dispossessed?

11. Fill in the blank with this word: "___-noir (modern film genre)"

12. Vatican vestment

14. The English translation for the french word: Celte

18. Fill in the blank with this word: "Costa del ___"

22. The English translation for the french word: daphnÈ

25. Ticket choice

26. Worry

27. The "E" of B.P.O.E.

29. Germany's Dortmund-___ Canal

31. Finish this popular saying: "No man is an_____."

32. Fill in the blank with this word: ""How ___ Has the Banshee Cried" (Thomas Moore poem)"

34. British tax

35. The English translation for the french word: rani

36. Aromatic Eurasian perennial

37. Fill in the blank with this word: "___ go bragh""

38. The English translation for the french word: sonnerie

39. U.S.N.A. grad

43. Limestone regions with deep fissures and sinkholes

44. The English translation for the french word: bÈvue

46. Flap raisers?

47. Port gets its name from this second-largest Portuguese city that's 2 letters longer

48. Unscramble this word: osenpr

50. The English translation for the french word: laÔcat

51. Holiday ___

52. Fill in the blank with this word: "___ bread"

55. Unpopular spots

56. Without a Trace' org.

57. Fill in the blank with this word: ""If the ___ is concealed, it succeeds": Ovid"

59. Tiny bit

PUZZLE 28

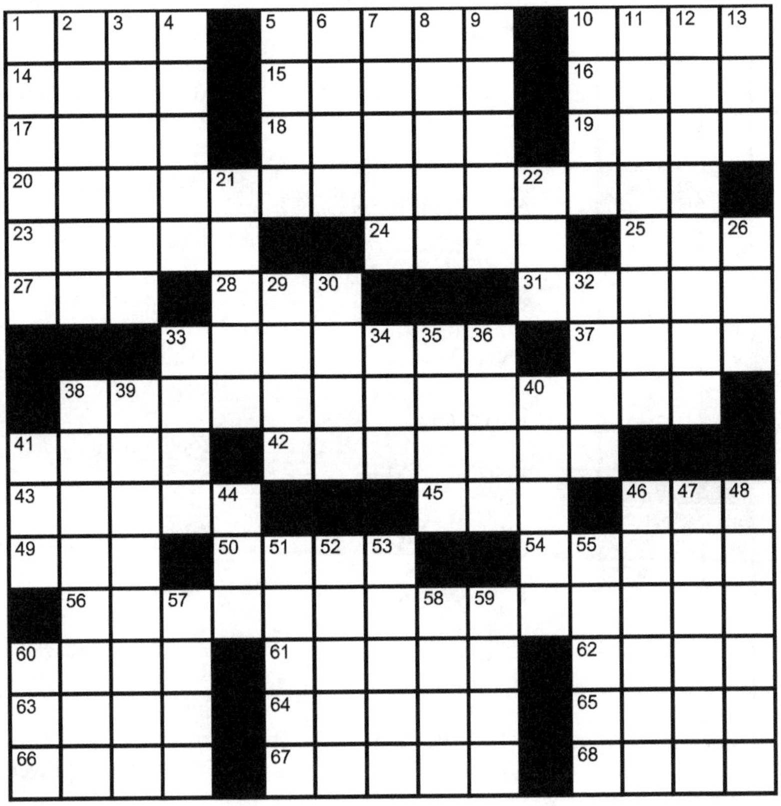

A crossword grid with numbered cells (1-68).

ACROSS

1. Aircraft carrier

5. Fill in the blank with this word: """___ flowing with milk and honey": Exodus"

10. Transcript stats

14. The Ponte Vecchio crosses it

15. Lethargy

16. Sound heard through a stethoscope

17. Where William the Conqueror is buried

18. The English translation for the french word: octal

19. Trapped like ___

20. Producer of many fragrances

23. Use the spade again

24. ___ weight

25. This is ___' (broadcast tagline)

27. Yellow ___

28. Winner of the first three Fiesta Bowls, for short

31. They're heard when Brits take off

33. Eskimo's environs

37. Song of India

38. Gift that almost killed Snow White

41. Sullen

42. "Of course!"

43. Fill in the blank with this word: "___ Chris Steak House"

45. Vapour trail?

46. Fill in the blank with this word: "___ Onassis, Jackie Kennedy's #2"

49. Fill in the blank with this word: "___ Digital Shorts (late-night comic bits)"

50. Org. for the Denver Gold and Chicago Blitz

54. Wearer of three stars: Abbr.

56. What the ends of 20-, 35- and 42-Across are, collectively

60. Fill in the blank with this word: """Mens sana in corpore ___""

61. Name in cosmetics since 1931

62. Writer Sarah ___ Jewett

63. Teen-___

64. Puts up

65. Critic, at times

66. Aristophanes work

67. Where le pr

68. Sound of a leak

DOWN

1. Stunning slaps

2. Town in County Kerry

3. Upstate New York's ___ Lake

4. Fill in the blank with this word: "___ scheme (investment scam)"

5. Tell ___ story

6. Track down

7. Is ___ (probably will)

8. Fill in the blank with this word: "___ Dove (the constellation Columba)"

9. 7 Faces of ___' (1964 film)

10. Sheepskin holder

11. relating to or characteristic of or befitting a parent

12. N.F.L. Hall-of-Famer elected to the Minnesota Supreme Court

13. Workout unit

21. They sustain many cultures

22. What's right in front of U

26. Mandela's land: Abbr.

29. Vessel with a load

30. The ___ Reader (magazine)

32. The sculptures "Cloud Shepherd" and "Coquille Crystals"

33. Seal's opening?

34. Drive forward

35. Year Attila was defeated in Gaul

36. Small islands

38. Weight

39. Sketched

40. Strips

41. Some Harvard grads: Abbr.

44. Fill in the blank with this word: """So ___ me!""

46. Town centers in old Greece

47. Participates in a class action

48. Puts under

51. Trauma aftereffects

52. U.S. Army training center in Va.

53. Fill in the blank with this word: "Biff ___, Arthur Miller character"

55. Fill in the blank with this word: "___ hammer (Viking symbol)"

57. Actor Calhoun

58. Fill in the blank with this word: "___ avis"

59. Way: Abbr.

60. Fill in the blank with this word: "Amniotic ___"

PUZZLE 29

ACROSS

1. Pampers rival

5. Used to entangle a cow's legs, gauchos make good use of this weapon of strong cords with weighted ends

9. Actor Sam

14. Step ___!'

15. When I Was ___' ('H.M.S. Pinafore' song)

16. Siouan speakers

17. Pirate's interjection

18. Fill in the blank with this word: ""Scrubs" co-star ___ Braff"

19. Wooly ruminant

20. Healthy

23. The English translation for the french word: Gustave

24. Uncompromising leader

28. Type of 35mm camera

29. Mancinelli's "___ e Leandro"

31. Wine valley

32. Belgian composer Guillaume

35. When lunch ends, maybe

37. Pitcher Robb ___

38. [See circles]

41. Fill in the blank with this word: "Dryden's "___ for Love""

42. Israel's Barak and Olmert

43. South American shrubs with potent leaves

44. Mardi Gras, e.g.: Abbr.

46. Washboard ___

47. Youth

48. Mouthed off

50. One with staying power?

53. Doesn't throw away, as a stage prop?

57. Girl's name that means "sorceress"

60. This Japanese photo film company outbid Kodak as a 1984 Olympics sponsor

61. Old laborer

62. Fill in the blank with this word: "___ Liebe (Dear, in Dresden)"

63. Fruit holder

64. I could ___ horse!'

65. 1988 Peter Allen musical

66. Pizzeria in Spike Lee's "Do the Right Thing"

67. U.K. military medals

DOWN

1. Vegetate

2. Strip, as a ship

3. Objets d'art

4. "Casablanca" villain

5. Place to buy a Persian rug

6. 1957-91 king of Norway

7. Peterson of 2003 news

8. Cause of some impulsive behavior, for short

9. Novelist Cather

10. Serious

11. Bygone carrier

12. When "S.N.L." wraps in N.Y.C.

13. Wiretapping grp.

21. Name on many a hospital

22. Words with line, hint or bomb

25. Stumbled upon

26. Fill in the blank with this word: "___ can of worms"

27. Relatives of "Gee whiz" and "Shucks!"

29. Writers Blyton and Bagnold

30. Wedding parties?: Abbr.

32. They test reasoning skills: Abbr.

33. Gettysburg general

34. Some greens

35. Day spa offering

36. Celtics head coach, 1995-97

39. You'll find it under a tree

40. Wish to a traveler

45. Cobra products

47. Superiors of sarges

49. Sniggled

50. Superman portrayer

51. Joy Adamson's "Forever Free: ___ Pride"

52. Non ___ (not so much, in music)

54. Some PX clientele

55. City rebuilt by Darius I

56. Sounds heard in passing?

57. T-shirt sizes, in short

58. Fill in the blank with this word: ""___-haw!""

59. Worth mentioning

PUZZLE 30

ACROSS

1. Savings acct. protector

5. The "H" in "M*A*S*H": Abbr.

9. Silkwood of "Silkwood"

14. TV star who shills for Electrolux

15. Well-running group?: Abbr.

16. Take ___ from someone's book

17. Sea lion, e.g.

19. Montana, e.g., once

20. With 52-Across, how to take some medications

21. PBS policy

23. When clocks are set ahead: Abbr.

24. Napoleonic marshal

25. Longtime New York senator for whom a center is named

28. Spain's Gulf of ___

29. Fill in the blank with this word: "___ and snee"

30. Dress up

33. Winery equipment

37. temperature measured by a mercury thermometer

38. Presided over

41. Turn on an axis

42. Parts of a baseball schedule

43. TD scorers

46. Fill in the blank with this word: ""The Suze ___ Show""

47. Annoying obligations / "No need to check" [split]

51. Valedictorian's pride: Abbr.

54. Southern ___

55. Musical notes

56. very large white gannet with black wing tips

58. Start of a refrain

60. Short witty remarks

62. Three-time Indy winner Bobby

63. Fill in the blank with this word: "Corn ___"

64. Succumb to mind control

65. Weigh station sights

66. Site of Ikea's hdqrs.

67. Dusseldorf donkey

DOWN

1. Now considered an environmental hazard, this refrigerant gas was discovered in the 1930s by Dr. Thomas Midgley

2. Steak ___

3. Supplication starter

4. Where William the Conqueror is buried

5. Heave-___ (dismissals)

6. Demilitarized place

7. Slimy shore deposit

8. Where many students click

9. Wyatt Earp, for one

10. Bee: Prefix

11. The Amazing ___ (noted magician)

12. Get rid of

13. "Shoot!"

18. Phenomenon of paramnesia

22. Wedding reception VIPs

26. Sixth Jewish month

27. Vance of "I Love Lucy"

28. Roman who originated the phrase "While there's life, there's hope"

30. They often accompany logos: Abbr.

31. Drive forward

32. Har-___ (tennis surface)

34. Usher's offering

35. Three-way joint

36. Vietnam War-era org.

38. Sweet potato nutrient

39. The English translation for the french word: hymne

40. Winner of all four grand slam titles

42. "Stop stalling!"

44. Bluish white twinklers

45. Word to Rover

47. Recurring metrical beat

48. French department

49. Germ ___ (chromosome locale)

50. Sporty car features

51. Unaccompanied part songs

52. Fill in the blank with this word: "___ que (because): Fr."

53. Wilderness photographer Adams

57. Fill in the blank with this word: "___ in a blue moon"

59. Wreath

61. Fill in the blank with this word: "___ Zeppelin"

PUZZLE 31

ACROSS

1. Kind of payment

6. Upstate N.Y. school

9. Stadium sounds

13. What lumberjacks leave

14. Fill in the blank with this word: "Faulkner's femme fatale ___ Varner"

16. Greenland base for many polar expeditions

17. 1932 #1 hit for Bing Crosby

18. Year Otto I became king of the Lombards

19. Gas: Prefix

20. Attractions near the Nile

23. Malay Peninsula's Isthmus of ___

24. Poor devil

25. Highway sight, formally?

31. Weightlifter types

32. Hgts.

33. On the ___ (fleeing)

36. Over there, poetically

37. The English translation for the french word: prose

38. Numerical prefix with oxide

39. Fill in the blank with this word: ""___ 'nuff!'"

40. Plant with two seed leaves

41. Tijuana lunches

42. Connected on only one side, as a town house

44. Regular care

47. Fill in the blank with this word: "___-Wan Kenobi"

48. Mad-dog singer?

53. Druggist for whom some commercial pills are named

54. Wide-mouthed pitcher

55. Paris's ___ Rivoli

58. You and who ___?' (fighting words)

59. Wrapped garment

60. Fill in the blank with this word: "Computer ___"

61. The supreme Supreme

62. WSJ competitor

63. Kind of sketch

DOWN

1. Woodstock supply

2. Fill in the blank with this word: "___ possidetis (as you possess, at law)"

3. a one-horse sleigh consisting of a box on runners

4. Performed brilliantly

5. Mark the beginning of

6. Prefix with angular

7. The English translation for the french word: pomper

8. Not well

9. Didn't just criticize

10. Wiped out, slangily

11. What's in carrots but not celery?

12. Fill in the blank with this word: "___ kebab"

15. Broadcasting unit?

21. Fill in the blank with this word: ""___ approved" (motel sign)"

22. Some airport data: Abbr.

25. The English translation for the french word: pourquoi

26. Fill in the blank with this word: "Architect ___ Ming Pei"

27. Middle of a run?

28. Fill in the blank with this word: ""___ beaucoup""

29. You have about 1 quart of this for every 30 pounds you weigh

30. Spanish direction

33. Solidarity's Walesa

34. Stub ___

35. Food flavoring brand

37. Teen's woe

38. Reserved

40. Title page?

41. a low stool in the shape of a drum

42. Fishing nets

43. Fill in the blank with this word: "Anderson's "High ___""

44. The English translation for the french word: pis

45. Film director Pier ___ Pasolini

46. Fermented drink made from rye bread

49. McGregor of "Trainspotting"

50. Weird: Var.

51. Glassmaking ingredient

52. Zeus' wife

56. Where ___ sign?'

57. Schubert's 'The ___-King'

PUZZLE 32

ACROSS

1. World Match Play Championship champ a record seven times

4. Fungal spore sacs

9. Fill in the blank with this word: "___ throat"

14. Texas ___

15. Thick-coated dog

16. Yeats's work

17. Fill in the blank with this word: "___ sponte (of its own accord, at law)"

18. Witty banquet figure

20. Bygone daily MTV series, informally

21. Wild Indonesian bovine

22. Relative of a 29-Down

23. Fill in the blank with this word: "___ "Le Morte d'Arthur""

25. Tribe in Manitoba

26. Flightless bird: Var.

27. Sci. course

28. Some undergrad degs

31. Short online message

33. Bees do it: Var.

35. Yves Klein found this heavenly color a symbol of pure spirit & made works that were just a field of it

36. What fun!'

37. Crooked

38. Classic name on stage

40. Suburb of Tokyo

41. Ways around: Abbr.

42. I could ___ horse!'

43. Bottom-of-letter abbr.

44. Seuss's "Horton Hears ___"

45. Start a voyage

48. Spanish hill

51. Start of a 1957 hit song

52. Magician's name ending

53. A pharaoh vis-

55. Computer units: Abbr.

56. The English translation for the french word: niche

57. Nothing runs like a ___' (ad slogan)

58. Self: Prefix

59. Whoopi's role in "The Color Purple"

60. Right turn ___

61. Vietnam War-era org.

DOWN

1. Kodak founder

2. Mrs. Bush

3. Predawn period

4. Semitic fertility goddess

5. The Earl of Sandwich, for one

6. Slangy greetings

7. Guy Lombardo's "___ Lonely Trail"

8. Pioneering anti-AIDS drug

9. Oh, please, that's enough'

10. We love ___ you smile' (old McDonald's slogan)

11. Softens in water, in a way

12. Go-aheads

13. Nickname for JosÈ

19. Unscramble this word: elrmey

24. Virginia's ___ River

25. Let's wax philosophical & wonder why Rousseau left only fragments of his opera about Daphnis & her

27. King ___ (dangerous snake)

28. Three-stringed instruments

29. Warehouse

30. Print tint

31. Trails off

32. About 40 degrees, for N.Y.C.

33. Unscramble this word: oopht

34. Workers with dogs, maybe

36. Warren Buffett, for one

39. English churchyard sight

40. Like neglected muscles

43. Winnie-the-Pooh's donkey friend

44. Japanese beer brand

45. Comparatively gritty

46. Preparing to bloom

47. Rolls

48. Pres., to the military

49. Combined, in Compi

50. O.T. book

51. Call in the game Battleship

54. Without further ___

PUZZLE 33

ACROSS

1. Carefully thought†out strategy

5. Nears retirement

9. Communion, e.g.

13. Agency concerned with civil aviation

14. Defendant's statement

15. Piano repairman, perhaps

16. Player who introduces the ball into a scrum

18. More likely to

19. Surveillance device

20. Word with first or financial

21. Fast follower?

23. High ground

25. Talked terms

27. ___ in the wool

28. Proper partner

29. Homer's neighbor

30. Fracas

33. Urgent prompting

36. The quality of affording easy familiarity and sociability

38. You need a cracker to eat it

40. Islamic sacred text (Var.)

41. What's put before the carte?

42. Send off, as broadcast waves

44. China item

48. Carefully thought†out strategy

51. Gracefully delicate

53. Second wife of Henry VIII

54. Trail the pack

55. Nonkosher food

56. Make scrimshaw

57. See 49-Across

60. Arab leader (Var.)

61. It's linked to mighty

62. Lust for life

63. Small amounts

64. Part of the eye

65. Sigmoid swimmers

DOWN

1. Thin†out buds

2. Absorb

3. Use mouthwash

4. A puzzle solver

5. Ladybug's lunch

6. Delighted

7. Japanese delicacy

8. Hunting journey, often

9. Indian money

10. Prisoner of war, e.g.

11. Native†American tent

12. Leave the straight and narrow

15. It may be plain or sweet

17. Supply with a staff

22. Gloomy

24. Billboard designer

25. Police matter

26. Insecticide banned since 1972

28. One-customer connector

31. Aussie avian

32. In the recent past

34. Romantic or Victorian, e.g.

35. Language producing "shampoo" and "pajamas"

36. Make loud demands

37. One of three in Fiji

38. Jokester

39. Chart-topping country band

43. China's last dynasty

45. Get some oxygen, e.g.

46. Pertaining to, or existing with reference to

47. A fold of tissue that partly covers the entrance to the vagina

49. North Pole artisans

50. Person of equal standing

51. Home on the Black Sea, perhaps

52. Of a previous time

54. Balcony area

56. Meow Mix consumer

58. It replicates in and kills the helper†T†cells

59. 'Enter the Dragon" star

PUZZLE 34

ACROSS

1. ___-Wan Kenobi
4. "That's a ___!"
8. British tax
12. "Dear" one
15. Animal house
16. Halo, e.g.
17. At, in or to any place
19. Twists together
21. Repeated by rote
22. Back
23. People or groups of people
25. ___ gestae
26. 100 qintars
27. Spotted, to Tweety
28. "___ Maria"
29. Petting zoo animal
32. "___ alive!"
34. Renaissance Fair props
36. Careers in general
39. A lapel on a woman's garment
40. Victorian, for one
41. Irritate
42. "___ we having fun yet?"
43. ___ terrier
45. A hand
46. Back talk
48. #1 spot
52. The "p" in m.p.g.
54. Like some humor
57. Ballad
58. Prompt
61. The customary or habitual hour
63. Bumper sticker word
64. Certain theater, for short
65. Brush up on
66. Befuddle
67. ___ v. Wade
69. Long, long time
72. Delivery vehicle
73. Visual property of something
76. Gibson, e.g.
79. Prepared leftovers
80. Gambling or speculation
82. Cover with a grit
83. Auspices
84. "My Name Is Asher ___" (Chaim Potok novel)
85. "C'___ la vie!"
86. Fill
87. Quaker's "you"
88. Clairvoyance, e.g.

DOWN

1. Horace volume
2. Bowed
3. Reciprocal action and reaction
4. Paneling for a low wall
5. Bleed
6. "He's ___ nowhere man" (Beatles lyric)
7. Antiquated
8. Crime boss
9. Passes
10. Aleppo's land
11. A person who sprawls
12. Gorge
13. Become covered with a layer of ice
14. Flush
18. Auction offering
20. Bring about
24. Small boat
29. Amazes
30. Caper
31. 27-Across, in dialect
33. Slippery substance
35. Eye layer
37. "Fantasy Island" prop
38. Silver ___ (photography compound)
39. Kind of wit
44. "Beowulf," e.g.
47. Setting for TV's "Newhart"
49. Customers collectively
50. "Oh, ___!"
51. Checked out
53. Petitions
55. Turn red or yellow, say
56. A deep fissure
58. Cast or model anew
59. Ceremony skippers
60. Blockbuster
62. Piece of land
64. Fresh
68. Old Roman port
70. Santa's reindeer, e.g.
71. Whinny
74. Affranchise
75. Halftime lead, e.g.
77. Assayers' stuff
78. Common request
81. Pandowdy, e.g.

PUZZLE 35

ACROSS

1. Yorick's skull, for one

5. Violent behavior, to Brits

10. Put a tree after "H" & you get this nautical wheel

14. Now see ___!'

15. Major defense contractor

16. Dilbert co-worker

17. Not delayed

19. Love ___

20. Le coeur a ___ raisons...': Pascal

21. Song that people flip for?

22. Witch of ___

23. Wiretapping grp.

24. Noted Presidential loser

26. Yeah, sorry'

30. Riviera resort

34. Prix de ___ de Triomphe (annual Paris horse race)

35. 1942-45 stats disseminator: Abbr.

37. Pal, in slang

38. Thin as ___

40. Female rap trio with the #1 hit "Waterfalls"

42. U.R.L. opener indicating an additional layer of encryption

43. The English translation for the french word: risquÈ

45. Watership down?

47. Yarn

48. State

50. Work on a galley

52. Shoe brand that sounds like a letter and a number

54. Mother ___

55. Ran over

58. Jewelry chain

60. Fill in the blank with this word: ""O Sole ___""

63. Ormandy's successor in Philadelphia

64. Old medicine?

66. See 103-Across

67. Syndicated astrologer Sydney

68. Western Indian

69. Fill in the blank with this word: ""___ Say," 1939 #1 Artie Shaw hit"

70. Worry greatly

71. Squire

DOWN

1. Letters from Greece

2. Oscar-winning French film director ___ Cl

3. ___ Island (location near Portland, Maine)

4. The Carolinas' ___ Dee River

5. Words starting a simple request

6. Partied, so to speak

7. Peer Gynt Suite' composer

8. Tough guys

9. Grand ___ Opry

10. Scribbled, e.g.

11. Dusseldorf donkey

12. Finish this popular saying: "He who hesitates is____."

13. Business school subj.

18. Made a tax valuation: Abbr.

22. -

23. [See blurb]

25. Unscramble this word: alp

26. Widen

27. Fill in the blank with this word: "___ to go"

28. You ___ right!'

29. Windsor's prov.

31. Fill in the blank with this word: ""Cũmo ___?""

32. Sugar ___

33. Word go

36. Big sizes, briefly

39. Kirsten of "Spider-Man"

41. Studio shout

44. Motel freebie

46. From memory

49. Skin condition

51. Younger brother, say

53. Suspect foul play

55. Struck, once

56. Actor Willard of "The Color Purple"

57. Spillane's '___ Jury'

59. Peseta : Spain :: ___ : Italy

60. Tiny bit

61. Windows picture

62. Kind of romance between actors

64. Stuff

65. Yankee Maris, informally

PUZZLE 36

ACROSS

1. Fill in the blank with this word: ""Beat ___ to ...""

6. Yarn

10. Massenet's "Le ___"

13. Stooge

14. 1960 Updike novel

15. Fill in the blank with this word: ""___," said Tom haltingly"

16. Booze

18. Wild ___

19. Lamb's "Essays of ___"

20. Wriggling

21. -

22. #1 best sellers

24. When "S.N.L." wraps in N.Y.C.

25. ... song by Billy Joel?

31. To-do

35. Maker of a dramatic 1971 getaway

36. Bone: Prefix

37. Repetition for rhetorical effect

39. Symphonie espagnole' composer

40. Phrase an overseas traveler should know how to translate

42. Fill in the blank with this word: "Cosmopolitan's ___ Gurley Brown"

43. Dual-purpose laundry room device

46. To his good friends thus wide I'll ___ my arms': 'Hamlet'

47. Quondam

52. World-weariness

55. Unit of speed

57. Olive genus

58. Sporty cars

59. They may have designs on you

61. Clinton's #2

62. Former Chinese Communist military leader Lin ___

63. Pippi Longstocking feature

64. U.S. ___

65. Fill in the blank with this word: "___-European"

66. Exceptional rating

DOWN

1. Fill in the blank with this word: "___ and pains"

2. The English translation for the french word: psaume

3. Via ___ (Roman road)

4. Fill in the blank with this word: "___ bar"

5. See 44-Across: Abbr.

6. Like best friends

7. Seed covering

8. Sitcom character discussed in the 2003 biography "Ball of Fire"

9. WSW's reverse

10. ___ latte

11. Whit

12. Word repeated in "Now ___ away! ___ away! ___ away ...!"

15. Unwelcome sight on an apple

17. Way off base?

21. Fill in the blank with this word: "___ palm"

23. Mao's successor as Chinese Communist leader

24. Fill in the blank with this word: "___ in a blue moon"

26. Staffordshire stench

27. Pugilists' org.

28. Flame Queen ___ (famous gemstone)

29. Ubangi tributary

30. Scientology's ___ Hubbard

31. Serious trouble

32. Workplace watchdog, for short

33. She-bears, south of the border

34. Sitting room?

37. The English translation for the french word: b˚cher

38. Israeli airport city

41. Visored cap

42. Realm of Otto I: Abbr.

44. It's not just me?'

45. Within: Prefix

48. Slaves

49. Characters in "Casablanca" and "Judge Dredd"

50. a half-breed of white and American Indian parentage

51. Bridge seats

52. Waf-fulls brand

53. Like many office jobs

54. Writer Ephron

55. ___ Capital (firm co-founded by Mitt Romney)

56. Very little

59. Object

60. Special ___

PUZZLE 37

ACROSS

1. Taken ___

4. Indian lute

9. Four hours on the job, perhaps

13. Fill in the blank with this word: "Baked ___"

15. Fill in the blank with this word: ""It's ___ against time""

16. Yes-___ question

17. Word: Suffix

18. Spreader of dirt

19. Sew up

20. Better than O.K.

22. Vassals

24. In tune

25. Pub orders

27. What a cedilla indicates

30. Reply facilitator: Abbr.

31. What a satellite may be in

35. Woodwind instrument: Abbr.

36. Actor Green and others

38. New Rochelle college

39. Neural network

40. Writer Joyce Carol ___

41. This Colette heroine is the granddaughter of a courtesan, who trains her to continue the family tradition

42. To stand in the center of this state, go 5 miles northeast of Ames & stand there--

27. What a cedilla indicates

43. "Borstal Boy" is the autobiography of this Irish author named Brendan

44. Coordinate in the game battleships

45. The English translation for the french word: psaume

47. Irish airline, ___ Lingus

48. Weapons check, in brief

49. Steel braces with right-angle bends

51. Without a Trace' org.

52. Native Hungarian

55. #5

60. Fill in the blank with this word: "___ helmet (safari wear)"

how exciting...

61. Fill in the blank with this word: "___, meenie, miney, mo"

63. Worrier's words

64. Answer to the riddle "Dressed in summer, naked in winter"

65. Fill in the blank with this word: "___ the mistletoe"

66. Hurricane-tracking agcy.

67. On ___ with (equal to)

68. Trappers' wares

69. Ball catcher

DOWN

1. Shirt name

2. Fill in the blank with this word: "___ Lamont ("Singin' in the Rain" role)"

3. Long dist.

4. Give the go-ahead

5. Mars: Prefix

6. Soul music over a financial institution's sound system?

7. Natl. Boss Day, ___ 16

8. Casino employee

9. Nears, as a target

10. Vestige

11. Katherine _____ Porter

12. Hebrew letters

14. Confessions of a drag queen?

21. Vitamin World competitor

23. Fill in the blank with this word: "___ Jima"

26. Try again

27. Temporary money

28. Vegetable fats

29. Islamic decree

30. Abalone

32. Nightclub

33. "Picnic" playwright's kin

34. Tarnish

36. Fill in the blank with this word: "___ sister"

37. W-2 info: Abbr.

46. Wharton grad

48. TV's "The ___ Today"

50. March on, march on, since we ___ in arms': Richard III

51. French town near Alen

52. Trailer org.?

53. Continental locales

54. Lack of vigor

56. Amusement park purchase

57. Fill in the blank with this word: "___ McAn shoes"

58. Fill in the blank with this word: ""While ___ it ...""

59. Banana ___

62. WSW's reverse

PUZZLE 38

```
 1   2   3   4       5   6   7   8       9   10  11  12  13
14              15                  16
17              18                  19
20              21              22  23
24          25                  26
        27              28  29  30      31  32  33  34
35  36  37          38              39      40
41              42                  43
44          45                  46
47          48      49              50
        51  52          53              54  55  56
57  58  59          60      61          62
63                  64  65          66
67                  68              69
70                  71              72
```

ACROSS

1. Muslim pilgrim

5. Water's conductivity comes from these particles, like positive sodium ones & negative chlorine ones

9. Wishes one can get on a PC?

14. Fill in the blank with this word: "Faulkner's "Requiem for ___""

15. Corp. money managers

16. Take care of the spread

17. Due

19. Suffix for wide-screen movie trademarks

20. Year in Claudius's reign

21. Whit

22. Place for a classic fight

24. The English translation for the french word: svelte

26. Fill in the blank with this word: ""Tais-___!" (French "Shut up!")"

27. Not free

31. Flame Queen ___ (famous gemstone)

35. ___ jazz (fusion genre)

38. Writer Asimov

40. Tail: Prefix

41. Colorful breakfast food

44. Meditation sounds

45. Fill in the blank with this word: ""To ___ human""

46. Like some private dets.

47. While you're enjoying the mountain scenery you might spot a tarn, a small one of these left by a glacier

49. Extreme Atkins diet credo

51. What secondary recipients of e-mails get

53. More pleased

57. Start to write?

61. Fill in the blank with this word: "___ el Amarna, Egypt"

62. Yugoslav novelist ___ Andric

63. Fill in the blank with this word: ""___ say it is good to fall": Whitman, "Song of Myself""

64. a material such as glass or porcelain with negligible electrical or thermal conductivity

66. You can bank on it

67. Scope

68. Society's dregs

69. What can be a real drag?

70. Fill in the blank with this word: "En ___ (all together)"

71. Unscramble this word: cpoy

72. Memory: Prefix

DOWN

1. Loudly commends

2. Cause for celebration: Abbr.

3. Musical London

4. Magician's name ending

5. Lemonade + ___ = Arnold Palmer

6. Fill in the blank with this word: "Easy as falling ___ log"

7. Emergency call

8. Ukr. and Lat., once

9. Fill in the blank with this word: "___-Lodge"

10. Part of many workouts, informally

11. Run up ___ (owe)

12. San ___, Italy

13. Bond villain in "Moonraker"

18. Opposite of destined

23. When clocks are changed back from D.S.T. in the fall

25. The Beatles' "Penny ___"

28. Meteorologist's prefix

29. "Hamlet" courtier

30. Fill in the blank with this word: "___ prayer"

32. Word after cream or powder

33. Fill in the blank with this word: "Bust ___ (laugh hard)"

34. TV's Anderson

35. Last name in ice cream

36. Longtime Vicki Lawrence character

37. Ear covering

39. Breaker?

42. Fill in the blank with this word: "007 foe ___ Blofeld"

43. False coin

48. Establishment of in a new environment, as a plant

50. Watery, as eyes

52. Big bill

54. Tool attached to a rope

55. The English translation for the french word: Èvoquer

56. Santa Fe Songs' composer

57. Gulf of ___, body of water next to Viet Nam

58. Fill in the blank with this word: ""___ Simple Man" (#1 Ricky Van Shelton song)"

59. Foreign attorneys' degs.

60. Superior setting: Abbr.

61. Encouraging sign

65. Operations ___ (Army position)

66. You can bank on it

PUZZLE 39

ACROSS

statement

1. Fill in the blank with this word: ""I Still See ___" ("Paint Your Wagon" tune)"

6. Therefore

10. Women, slangily

14. The English translation for the french word: BÈnin

15. Very recently: Abbr.

16. Zany Martha

17. The Green Hornet's real first name

18. ___ latte

19. French wave

20. 1986 Prince movie, after 29-Down

23. Muhammad ___

25. Operations ___ (Army position)

26. Volunteer's

27. Designer Versace

29. Fill in the blank with this word: "___ Kea"

32. Helmetlike flower petal

33. Turns sharply

34. Romanian money

37. A restaurant patron said ...

41. Year that Augustus exiled Ovid

42. Some military defenses, for short

43. Gravelly ridge

44. Truth ___

46. What many villains come to

47. Wolf disguised as one of these gets killed by shepherd wanting dinner

50. Singer ___ King Cole

51. Fill in the blank with this word: ""___ approved" (motel sign)"

52. Gets lucky

57. 1981 hit film with a 5'3" lead actor

58. Fill in the blank with this word: "___ bird"

59. [See title, and proceed]

62. O'er bank and ___ ... he glanced away...': Sir Walter Scott

63. Spiritedly, in music: Abbr.

64. Old lab heaters

65. Way: Abbr.

66. Butt out,' briefly

67. The English translation for the french word: orteil

DOWN

1. Withdraw

2. Celtic sea god

3. Okayed, in a way

4. Villainous group in the 'Star Wars' universe

5. These segmented sensory appendages on the heads of insects are thought to be touch & smell receptors

6. Danish astronomer Brahe

7. Titter

8. Org. for part-time soldiers

9. Fill in the blank with this word: "___-fry"

10. See 33-Across

11. What las novelas are written in

12. Fill in the blank with this word: ""___ has fleas" (uke-tuning phrase)"

13. Fill in the blank with this word: ""I ___ reason not to""

21. Year the emperor Decius was born

22. You're in balance if you know it's the complement of "yang"

23. Texas A & M student

24. Tropical vine

28. Wheeler Peak locale: Abbr.

29. Founding member of the Washington Freedom

30. Ticket sellers: Abbr.

31. Fill in the blank with this word: "___ Ronald Reagan"

33. Indian bovine

34. Extravagant way to live

35. Fill in the blank with this word: ""Maria ___," Jimmy Dorsey #1 hit"

36. Std. on food labels

38. Soldier

39. Spruced up

40. Comparable in years

44. Unscramble this word: rsecte

45. N.T. book

46. Fill in the blank with this word: ""A guy walks into a ___ "

47. Signs of healing

48. Plague

49. Singers James and Jones

50. Certain W.M.D.

53. Fill in the blank with this word: "___ cheese"

54. Fill in the blank with this word: "___ Pictures"

55. Sound

56. Palio di ___ (Italian horse race)

60. Fill in the blank with this word: "Bao ___ (former Vietnamese emperor)"

61. Fill in the blank with this word: ""Humanum ___ errare""

PUZZLE 40

ACROSS

1. Like some presidents

7. Words with a nod

11. Wildlife threat, briefly

14. The English translation for the french word: lipase

15. Fill in the blank with this word: "Explorer ___ Anders Hedin"

16. Pupil's place

17. Unwilling (to)

18. Flip

19. Some clergy: Abbr.

20. Wiser from an ethical perspective?

23. Jay-Z's ___-Fella Records

26. Morse tap

27. She, in Italy

28. They may grant immunity

31. Rink star Phil, to fans

34. Fill in the blank with this word: "___-noir (modern film genre)"

35. Rice-A-___

37. The English translation for the french word: cinÈma

41. Get kicked?

44. Some natural history museum displays, for short

45. Fill in the blank with this word: "___ Adams, signature on the Declaration of Independence"

46. What a lei person might pick?

47. Fill in the blank with this word: "___-Ude (Trans-Siberian Railroad city)"

49. Spanish inns

51. Jay who once hosted "Last Comic Standing"

54. Mount ___, active Philippine volcano

56. George Harrison's "___ It a Pity"

57. 34-Down for a bookkeeper?

62. HVAC measure

63. Wanton look

64. Start of a presidential march

68. Verb type: Abbr.

69. The English translation for the french word: berme

70. Time competitor, for short

71. Fill in the blank with this word: "Brig. ___"

72. Golf innovator Callaway and bridge maven Culbertson

73. Palooka

DOWN

1. Mao's mil. force

2. Fill in the blank with this word: ""6 Rms ___ Vu" (play)"

3. To his good friends thus wide I'll ___ my arms': 'Hamlet'

4. Touchdown setting

5. Fill in the blank with this word: "___ Plaza, former Calgary landmark name"

6. Physics Nobelist Simon van der ___

7. Pancreatic hormone

8. The English translation for the french word: svelte

9. Start of a decision-making process

10. Finish this popular saying: "Every stick has two___."

11. Fill in the blank with this word: "Cheese ___ (snack)"

12. Goddess of agriculture

13. 'The Venice of the Middle East'

21. John Hersey's 'A Bell for ___'

22. Time without end

23. Title role in a 1950s TV western

24. Two semesters

25. Singer Sam

29. One of the archangels

30. Fill in the blank with this word: "Dom DeLuise sitcom "___ Luck""

32. Stupid jerk

33. Pabst brand

36. Fill in the blank with this word: "___ and outs"

38. Israel's Barak and Olmert

39. 1949-51 N.B.A. top scorer George

40. Fill in the blank with this word: "Emily Dickinson poem "For Every Bird ___""

42. Part of a metropolitan region

43. Strike lightly

48. Tablecloths and such

50. Bedridden, say

51. Top banana

52. Weird

53. Rolaids target

55. Salon works

58. Wittenberg's river

59. Whirl

60. Thick-bodied fish

61. Old capital of Romania

65. Untilled tract

66. Local political div.

67. Verb ending

PUZZLE 41

ACROSS

1. Fill in the blank with this word: ""Life ___ a dream""

6. You might take investing tips from this network's "On the Money" or "The Call"; Jon Stewart probably doesn't

10. Druggist for whom some commercial pills are named

14. Saturday Night Live' alum Cheri

15. Start with pad or port

16. Not ___ many words

17. Praying figure

18. Vivacious wit

19. The Farmer in the ___'

20. It appears first in China

22. Here ___, there...' ('Old MacDonald' lyric)

23. Wedding page word

24. Possibly

26. Lux. neighbor

29. Pouches

32. Actress Skye and others

33. Slo-___ fuse

34. Central knob of a shield

35. Staten Isl., for one

36. Buying guides?

42. Wood of the Rolling Stones

43. Dr.'s orders

44. Talking-___ (scoldings)

45. Fill in the blank with this word: "___ positive"

48. Urge

49. Unit of loudness

50. Short-hop plane

52. One in the charge of un instituteur

54. Sari-clad royal

55. Where the African Union is headquartered

61. Short and disconnected: Abbr.

62. Liquid ___ (refrigerant)

63. Structural support

64. I could ___ horse!'

65. Fill in the blank with this word: "___ death (overwork)"

66. Symbol of a lingering scandal

67. Dusseldorf donkey

68. Santa ___ (neighbor of Lompoc, California)

69. Fill in the blank with this word: "___ message"

DOWN

1. Benevolent fraternal soc. since 1819

2. Purchase from a jeweler

3. You'll smile big if you know this 4-letter word for a long, thick piece of lumber used to support a roof

4. Entombing, old-style

5. Threepeater's threepeat

6. Julie of "The Early Show"

7. Tiber tributary whose name means "black"

8. They may pull banners

9. One quoted

10. Went around in circles?

11. Like some shows

12. Tilted

13. "It won't be missed"

21. The English translation for the french word: bÈni-oui-oui

25. Unrest

26. Toy gun shot

27. She, in S

28. You're killing me,' textually

30. Washboard ___

31. Fill in the blank with this word: "___ Girl"

34. Uncle Sam's land

35. They're rarely hits

37. Blue

38. Saccharin discoverer ___ Remsen

39. To the ___ power

40. Fill in the blank with this word: ""___ bad!""

41. W-2 info: Abbr.

45. Minority member in India

46. Vaquero gear

47. Victorian, in a way

48. One of the judges in Judges

49. Little rock

51. 2005 #1 album for Coldplay

53. Flock members

56. The English translation for the french word: dorloter

57. Fill in the blank with this word: ""Rinkitink ___" (L. Frank Baum book)"

58. Fill in the blank with this word: ""___ Flux" (Charlize Theron film)"

59. Unit of speed

60. Katharine's role in "Adam's Rib"

PUZZLE 42

ACROSS

1. Wax insert, perhaps

5. Very smooth

9. Window type

14. Mythical king of the Huns

15. Nary ___

16. Unsmooth

17. Call in the game Battleship

18. The Force was with him

19. Pre-Little League game

20. Saturated

23. Warbler Sumac

24. Fill in the blank with this word: "___ Friday's"

25. Sumter and McHenry: Abbr.

28. One featured in una galer

32. Kenyan president Daniel arap ___

35. Finish this popular saying: "You can have too much of a good___."

37. Working in a mess

38. Lustrous velvet

39. BOHR

42. Spanish noblewoman ___ de Castro

43. Unusual shoe spec

44. Fly catcher

45. Turkey, to a bowler

46. Words often said while kneeling

48. Union ___: Abbr.

49. Toni Morrison's "___ Baby"

50. War on Poverty agcy.

52. Delicacy

61. Threesome

62. Saint Philip ___ (Renaissance figure)

63. What's going ___ there?'

64. Sister of the Biography Channel

65. Saxophonist Zoot

66. Joined at the altar

67. Wood finish

68. Fill in the blank with this word: "___ du jour"

69. Wild goose

DOWN

1. Bankrolls

2. What's ___ you?'

3. See 26-Across

4. Tightly curled

5. Special intuition, in modern lingo

6. Some thieves in the night sought the troubadour's ___; they found not a dime, just a 6-stringed ___

7. Fill in the blank with this word: "___-European"

8. Vanquished

9. Spanish essayist ___ y Gasset

10. Yom Kippur service leader

11. Golf's ___ Aoki

12. One-named singer/actress

13. Singer Lovett

21. Slate and Salon

22. U.R.L. opener indicating an additional layer of encryption

25. N.J. post

26. "I appreciate it," in text messages

27. Vicks spray brand

29. W. C. Fields persona

30. Fill in the blank with this word: ""And to those thorns that ___ bosom lodge": Shak."

31. Brilliantly blue

32. Yum-Yum, Peep-Bo and Pitti-Sing in "The Mikado"

33. Word go

34. The English translation for the french word: interne

36. Sue Grafton's '___ for Noose'

38. Narc's find, for short

40. Finish this popular saying: "Cold hands, warm___."

41. Fill in the blank with this word: ""___ little silhouetto of a man" ("Bohemian Rhapsody" lyric)"

46. Word with aunt or voyage

47. Single-minded theorizer

49. Letter before qoph

51. Boy band with the hit "Liquid Dreams"

52. Union and others: Abbr.

53. Quod ___ faciendum

54. Singer Simone

55. Police dept. employee

56. Wedding wear

57. Fill in the blank with this word: ""Forever, ___" (1996 humor book)"

58. Sell short

59. Nothing, in Nice

60. The Neverending Story' author

PUZZLE 43

ACROSS

1. Hornet, e.g.

6. Defeat

11. What secondary recipients of e-mails get

14. Sister of the Biography Channel

15. Terse

16. Fill in the blank with this word: ""Am ___ risk?""

17. pertaining to time-honored orthodox doctrines

19. Har-___ (tennis surface)

20. Tried to tackle, say

21. Fill in the blank with this word: ""___, My God, to Thee""

23. Switzerland's ___ Leman

25. Fill in the blank with this word: "Chicago's ___ Expressway"

28. Unwanted buildup

29. Fill in the blank with this word: ""What's Hecuba to him ___ to Hecuba": Hamlet"

31. Blue, say

34. Snail trail

36. This city already had a bad reputation when Lot decided to settle there

37. Ben Jonson poem

40. Egg holders

44. The English translation for the french word: rÈbus

46. Pope's "An Essay ___"

47. Gen. Hooker fought in it

52. Fill in the blank with this word: "___ bag"

53. Ravel's "Gaspard de la ___"

54. "Pride and Prejudice" beau

56. Verb ending

57. It's a piece of cake

60. Ward off

62. My ___, Vietnam

63. MINED

68. Parisian article

69. This garlic-flavored mayonnaise from Provence is popularly served with fish

70. Pipsqueak

71. Whirlpool alternatives

72. Opposite of dimin.

73. The English translation for the french word: laÓche

DOWN

1. Singer ___ King Cole

2. Fill in the blank with this word: ""A guy walks into a ___ "

3. Lawless

4. Work of prose or poetry

5. Marie Antoinette, e.g.

6. Treated with malice

7. Thumbnail item

8. When some stores open

9. The Chinese Parrot' sleuth

10. MacLachlan of "Twin Peaks"

11. Lemonlike fruit

12. Steve of "The Office"

13. Well-built

18. Word of mock fanfare

22. What you may have while solving this puzzle?

23. Finish this popular saying: "He who hesitates is ___."

24. Fill in the blank with this word: "Comic strip "___ & Janis""

26. Xi preceders

27. Untidy one

30. They're no longer active

32. Y. A. Tittle scores

33. Mozart's "Dove ___"

35. Your highness?: Abbr.

38. There: Lat.

39. Fill in the blank with this word: ""___ Lang Syne""

41. Sandpapered

42. Uses a shuttle

43. Snick-or-___

45. United competitor: Abbr.

47. Remove from power?

48. Actress Langdon

49. They have their limits

50. With French, one of two official languages of Chad

51. Stereo syst. component

55. Cries of pain

58. Yao Ming teammate, to fans

59. ___ to middling

61. Way from Syracuse, N.Y., to Harrisburg, Pa.

64. Shad ___

65. Big sizes, briefly

66. School subj.

67. Ham on ___

PUZZLE 44

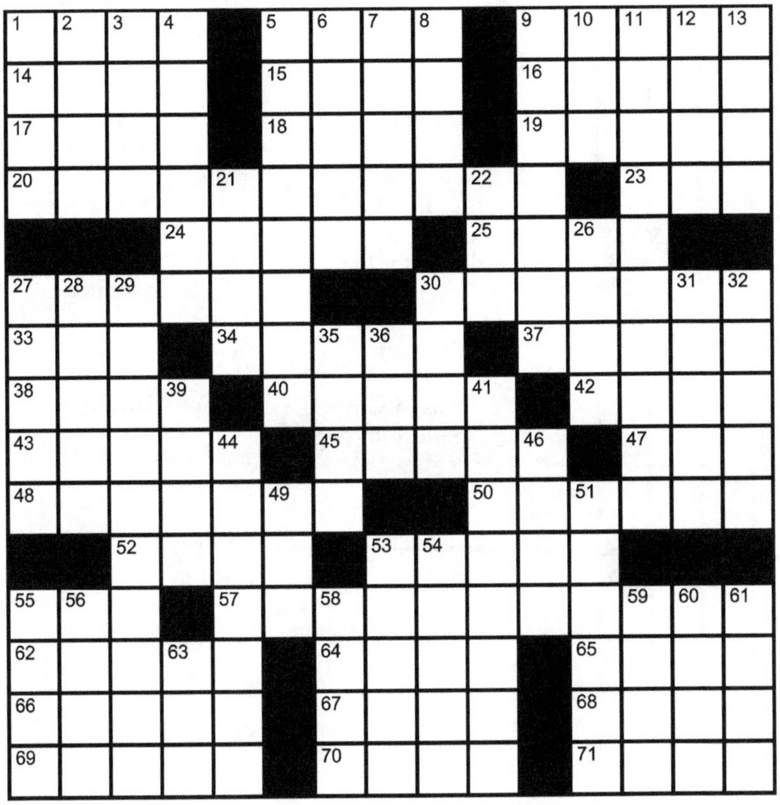

ACROSS

1. Where change is made

5. Word said just before opening the eyes

9. Where to live the high life?

14. Verb-to-noun suffix

15. Reindeer-herding people

16. Their ranks don't include DHs

17. Garage container

18. Takes some courses?

19. past (similar term)

20. Cast events after filming is done

23. Verb ending

24. Unscramble this word: yssea

25. Onetime big inits. in car financing

27. Some aromatic resins

30. Went into first, maybe

33. Popular cable channel

34. Kid's name

37. Let ___

38. Wield a mop

40. Western deal since 1994: Abbr.

42. Last name in ice cream

43. Wiesbaden's state

45. The second part missing in the author's name ___ Vargas ___

47. Jolly old ___ (Santa)

48. Magazine department

50. Yen

52. Wide-mouthed pitcher

53. They're S-shaped

55. Robert Louis Stevenson's "___ Triplex"

57. Fill in the blank with this word: "... Mrs. Herr decided she'd ___ albums"

62. Knows the answer

64. Year Attila was defeated in Gaul

65. Place for a Christmas card

66. Shed ___

67. Sound system brand

68. Ici ___ (here and there, to Th

69. Fill in the blank with this word: "___ profundo"

70. Ural River city

71. Fill in the blank with this word: "___ canal"

DOWN

1. Fill in the blank with this word: "___ Mix"

2. Go up: Abbr.

3. Wildcats' org.

4. Many an agent

5. Offer?

6. Spanish actress Carmen ___

7. void (similar term)

8. Fill in the blank with this word: "___ prius (trial court)"

9. Inability to smell

10. Wright wing?

11. Submarine in a Tom Clancy best seller

12. Ticks off

13. Old laborer

21. Sean Connery, for one

22. French vineyard

26. Fiery

27. Sister in "Three Sisters"

28. Swung, nautically

29. Gets a whole new view of

30. Fill in the blank with this word: "Explorer Hernando de ___"

31. Organic compounds

32. Vermeer's home

35. Pizzeria in Spike Lee's "Do the Right Thing"

36. Toronto Argonauts' org.

39. Wood-cutting tool

41. *Sign to look elsewhere

44. Villain of Spider-Man

46. Expressed wonder

49. We'll teach you to drink deep ___ you depart': Hamlet

51. The English translation for the french word: isomÈre

53. On in years

54. Dos

55. Wilt thou not chase the white whale?' speaker

56. I could ___ horse!'

58. Time-share unit

59. Germany's ___ von Bismarck

60. Realtor's specialty, for short

61. Wooden piece

63. Fill in the blank with this word: "___ in ink"

PUZZLE 45

ACROSS

1. Mt. Apo's locale, in the Philippines

6. Adjust, as a clock

10. Old Fords

14. Words said on the way out the door

15. Ziegfeld Follies designer

16. Answer to the riddle "Dressed in summer, naked in winter"

17. Pennsylvania Avenue

20. Launch ___

21. Field worker

22. She's a hip-hop fan

23. Filmmaker Riefenstahl

25. Fill in the blank with this word:

"Explorer Cabeza de ___"

27. Octopus, e.g.

31. Fill in the blank with this word: "45 ___"

34. Every, to a pharmacist

35. Hydrocarbon suffixes

36. Fill in the blank with this word: "___ special"

39. Nymph chaser

41. Genealogist's abbr.

43. Winter Palace residents

44. Princess Fiona, for one

46. Smartphone introduced in 2002

48. Writer Marilyn ___ Savant

49. World's first carrier with a transpolar route

50. Hard drives?

53. Lemon ___

55. Formal hat, informally

56. Singer K. T. ___

59. Whitish

61. Within: Prefix

65. Shortly after quitting time, for many

68. Fill in the blank with this word: "___-dieu"

69. Sound stressed, maybe

70. Like Miss Muffet's fare

71. Now see ___!'

72. Shoe part

73. How can ___?'

DOWN

1. Business slumps

2. To me, to Mimi

3. The English translation for the french word: volt

4. Off the mark

5. Thomas Moore poem "___ in the Stilly Night"

6. Sprang back

7. Upstate New York's ___ Canal

8. Fill in the blank with this word: ""... ___ a fever""

9. They dug his grave ___ where he lay': Sir Walter Scott

10. Slugs

11. U.S.S. Enterprise counselor

12. Fill in the blank with this word: "___ John"

13. Unscramble this word: lsle

18. Fill in the blank with this word: ""Wie geht es ___?" (German greeting)"

19. Year that Dionysius of Halicarnassus is believed to have died

24. They hold water

26. Fill in the blank with this word: ""If the ___ is concealed, it succeeds": Ovid"

27. Unexciting

28. Slate and Salon

29. Bony cavities, anatomically

30. They test reasoning skills: Abbr.

32. Fill in the blank with this word: "___ Nurmi, the Flying Finn"

33. Ancient Greek sculptor famous for his athletes in bronze

37. Milk: Prefix

38. Sound of a leak

40. Emphatic agreement

42. Skateboard wheel material

45. Spain's Costa del ___

47. TV sports awards

51. Relate to

52. Angry

54. Windsor's prov.

56. Not fer

57. Finish this popular saying: "Slow but___."

58. Fill in the blank with this word: "Dragon's ___ (early video game)"

60. Hydros : England :: ___ : U.S.

62. Fill in the blank with this word: "De ___ (Nolte's "Cape Fear" co-star)"

63. VHS tape displacers

64. Cyclops' feature

66. Fill in the blank with this word: "Capitol-___ (music company)"

67. Cable TV giant

PUZZLE 46

ACROSS

1. Fill in the blank with this word: "___ on (orders to attack)"

5. Mr. ___ of "Peter Pan"

9. Schindler of "Schindler's List"

14. Fill in the blank with this word: "En ___ tiempo (formerly, to Felipe)"

15. Means ___ end

16. Actor Raf of "The Italian Job," 1969

17. Stereotypically smarmy sorts

20. Tracy's "Tortilla Flat" co-star

21. the earth from his flesh

22. Verdi's "___ giardin del bello"

23. Union ___: Abbr.

24. The English translation for the french word: ‡ma

26. Chase of "Now, Voyager"

28. Steel braces with right-angle bends

30. European capital

34. Where D.D.E. went to sch.

37. Unscramble this word: pkac

39. Statehouse official: Abbr.

40. [See circles]

43. Spanish festival

44. High-tech transmission

45. Actress Mary et al.

46. This breathing technique used in childbirth is named for a French doctor

48. Fill in the blank with this word: ""It Had to Be You" composer ___ Jones"

50. Fill in the blank with this word: ""L'Shana ___ " (Rosh Hashanah greeting)"

52. Uncle ___

53. Fill in the blank with this word: ""Tais-___!" (French "Shut up!")"

56. They line up between centers and tackles: Abbr.

59. Like some high-fiber cereal

61. What the weary get, so they say

63. Czar to those who knew him when

66. Singer Lopez

67. Illustrator Silverstein

68. Fill in the blank with this word: "___-fry"

69. Winning craps roll, in Rome

70. Weight not charged for

71. Pouches

DOWN

1. People

2. Where ___

3. Espresso topping

4. Scotch and ___

5. Some Beverly Hills tourist purchases

6. W. Sahara neighbor

7. user-friendly (similar term)

8. As a friend, to the French

9. It's too much

10. World's first carrier with a transpolar route

11. J train?

12. You'll use up 3 vowels playing this word that means toward the side of a ship that's sheltered from the wind

13. Rice-A-___

18. Wrongfully take

19. Vocal rise and fall

25. Fill in the blank with this word: ""Laborare est ___ " ("to work is to pray")"

27. Sketcher's eraser

28. Western ring

29. Wells works

31. Where bacteria may flourish

32. Second to ___

33. Stock market figs.

34. Org. for the Denver Gold and Chicago Blitz

35. Where the Mets play

36. School ___

38. Fermented drink made from rye bread

41. Lilliputian of early 50's TV

42. What Bunyan grasped

47. Fill in the blank with this word: "Cole Porter's "Well, Did You ___?""

49. To me, to Mimi

51. This is only ___'

53. Anatomical roofs

54. "Hamlet" courtier

55. Anatomical canals

56. What sgts. turn in at HQ's

57. Ryder's "Autumn in New York" costar

58. "Cease and desist!"

60. Jewish youth org.

62. The supreme Supreme

64. Treebeard, e.g.

65. Most miserable hour that ___ time saw': Lady Capulet

PUZZLE 47

ACROSS

1. Letterman airer

6. Unusually excellent

10. Fill in the blank with this word: "Central Africa's Lake ___"

14. The English translation for the french word: rumba

15. Fill in the blank with this word: ""The very ___!""

16. Fill in the blank with this word: "Europe's Gorge of the ___"

17. What ___!' ('Bummer!')

18. P.O. box, e.g.

19. Fill in the blank with this word: "Ars longa, ___ brevis"

20. Misery

21. Rich Little

24. Barefaced

25. Projected onto a screen

26. 35-Across in 1994

31. Tool for some group mailings

32. Wood: Prefix

33. Fill in the blank with this word: "___ Perignon"

36. Fill in the blank with this word: "___ one"

37. Toiletry item

39. Honi __ qui mal y pense

40. Source of some rings

41. Greek goddess Athena ___

42. Popular "jam band"

43. Folded-up salary check?

46. Hero of a John Irving best seller

49. Weather info: Abbr.

50. Cast events after filming is done

53. What's funded by FICA, for short

56. Woman's name suffix

57. Silent parts of 20-, 36- and 49-Across

58. Hero of the 1997 best seller "Cold Mountain"

60. Critic, at times

61. Tombstone name

62. Tarantula-eating animal

63. Swedish actress Persson

64. Actress Schneider

65. Fill in the blank with this word:

DOWN

1. To study intensely at the last minute for a test

2. Chums

3. Kids' TV character

4. TV schedule abbr.

5. Hobo

6. Venice tourist attraction

7. Supplementary: Abbr.

8. Fill in the blank with this word: "___-Tea (first instant iced tea)"

9. Salary

10. Triangular toast topping

11. Fill in the blank with this word: "Actor ___ S. Ngor"

12. Musical Shaw

13. Ruthlessly competitive

22. Without a prescription: Abbr.

23. Fill in the blank with this word: "___ Detective (1930's-50's crime fiction magazine)"

24. Tower site

26. Numbers game

27. Wells's oppressed race

28. The head Corleone

29. Fill in the blank with this word: "___-Kettering Institute"

30. Word repeated in Emily Dickinson's "___ so much joy! ___ so much joy!"

""Smoking ___?""

33. The Everly Brothers' "All I Have to ___ Dream"

34. Fill in the blank with this word: "Auvers-sur-___, last home of Vincent van Gogh"

35. That you should feed a cold and starve a fever, and others

37. The English translation for the french word: papier tue-mouches

38. Wreath

39. "Big deal"

41. ___ The Magazine (bimonthly with 35+ million readers)

42. Its brands include Frito-Lay and Tropicana

43. Tissuelike

44. Fried chicken choice

45. Singer Des'___

46. Quaint contraction

47. Spanish counterparts of mlles.

48. Slalom targets

51. Island in French Polynesia

52. Fill in the blank with this word: "___ limits (election issue)"

53. One on a mission, maybe

54. Fill in the blank with this word: "1974 Peace Nobelist Eisaku ___"

55. Have ___ with (know well)

59. Fill in the blank with this word: "___'wester"

PUZZLE 48

ACROSS

1. Arthur and others

5. To ___ is human ...'

8. Fighting: 2 wds.

13. Vision: Prefix

14. Titter

15. Nirvana attainer

16. Acted like a baby, in a way

18. Fill in the blank with this word: "___ incognita"

19. Spiral-horned African antelope

20. Like ocher

22. Whitman's "A Backward Glance ___ Travel'd Roads"

24. Fill in the blank with this word: ""O Sole ___""

25. Murder mystery setting, to a cowboy?

33. Witch's ride

34. Some of a Beatles refrain

35. To annoy, perhaps with an insect

36. Sketch

37. Fill in the blank with this word: ""___ it from me...""

38. Barnum midget

39. Fill in the blank with this word: ""___ was saying Ö""

40. XM ___

41. Some urban digs

42. "Tough!"

45. Tiny energy units, for short

46. Nasdaq unit: Abbr.

47. Troubadour's creation

52. What las novelas are written in

56. Lord, ___?'

57. Parade V.I.P.

60. Split

61. Fill in the blank with this word: "___ splicing"

62. Tracy

63. Corporate shuffling, for short

64. Yacht's dir.

65. Ming's 7'6" and Bryant's 6'6", e.g.: Abbr.

DOWN

1. Team on which Larry Bird played, on scoreboards

2. One-named singer/ actress

3. That's not ___!' (parent's warning)

4. Japanese noodles, whether udon or these buckwheat-flour ones, are often served in broth

5. Most miserable hour that ___ time saw': Lady Capulet

6. Stat starter

7. "Little" barnyard bird with an alliterative name in a classic Willie Dixon blues song

8. Plants with aromatic oils

9. Botanist Mendel

10. Fill in the blank with this word: ""___ the brinded cat hath mew'd": "Macbeth""

11. Fill in the blank with this word: ""Star ___""

12. Fill in the blank with this word: "___ Beach (D-Day site)"

14. Gangster's weapon

17. Leopold ___, 'Ulysses' protagonist

21. Fill in the blank with this word: ""I ___ Walrus""

23. What is the capital of this country - Saudi Arabia

25. Actress Van Devere

26. Good friend, informally

27. Years on end

28. Weird

29. Wheels

30. Text on an iPad, say

31. Water pits

32. Three-stripers: Abbr.

33. Unappealing trumpet sound

37. String bean's opposite

38. Shelley's "___ Skylark"

40. Answering

41. Indonesian island

43. Limo driver in the airport, e.g.

44. Part of a sentence, in linguistics

47. Way to get to N.Y.C.

48. Toddler's attire

49. In ___ (type of fertilization)

50. Fill in the blank with this word: "Competitive ___"

51. They replaced C rations

53. Sodium hydroxide, to chemists

54. Go belly-up

55. Yahoo

58. Parisian article

59. Baseball positions: Abbr.

PUZZLE 49

ACROSS

1. Xanadu river

5. 1988 Peter Allen musical

10. The English translation for the french word: Èpaulard

14. Fill in the blank with this word: "Aloe ___"

15. With 52-Across, how to take some medications

16. Sketch

17. Year in the reign of Edward the Elder

18. Up ___ (trapped)

19. June 6 1944 is____.

20. Modeling requirement?

23. Jesse who lost to Ronald Reagan in 1970

24. Word derived from Japanese for "empty orchestra"

28. See 85-Across

29. Kupcinet and Cross

33. San ___, locale just north of Tijuana, Mexico

34. Whups

36. Pork ___

37. Components of some auto engines

41. Way to get to N.Y.C.

42. Vaporize

43. Not unless

46. New York's Carnegie ___

47. Word modifier: Abbr.

50. Doesn't stop

52. Female vampire

54. See 23-Across

58. You've Really ___ Hold on Me'

61. Symbol of generosity

62. Serb, e.g.

63. Norse goddess of fate

64. Yellow, perhaps

65. Yeats's land

66. Show ___ (attend, as a meeting)

67. Entry in a spaceship log

68. Zaire's Mobutu ___ Seko

DOWN

1. Tech-savvy school grp.

2. Tart

3. Record listings

4. The English translation for the french word: haÔku

5. Vegetate

6. Within: Prefix

7. Teri ___, Best Supporting Actress nominee for 'Tootsie'

8. Fill in the blank with this word: "___ preview"

9. Wedding hiree

10. Noted London landmark

11. Unscramble this word: dri

12. Org. with an annual televised awards ceremony

13. Fill in the blank with this word: "At ___ rate"

21. Bake, as eggs

22. Surgery sites, for short

25. Whiff

26. Fill in the blank with this word: "___ Kringle"

27. Years on end

30. Fill in the blank with this word: "___ room"

31. Fill in the blank with this word: "___ con Dios (Spanish farewell)"

32. Favored, with "on"

34. Agave

35. Your majesty'

37. Fill in the blank with this word: ""___ by me""

38. The English translation for the french word: orle

39. Next to ___

40. See 22-Down

41. Fill in the blank with this word: "___ Sabha (Indian legislature)"

44. Uniform: Prefix

45. Party items

47. Norwegian novelist/ feminist ___ Skram

48. Mideast dough

49. Freshmen's and sophomores' team

51. Warmish

53. Vaulted areas

55. Print tint

56. Site for techies

57. Fill in the blank with this word: ""Die Frau ___ Schatten" (Strauss opera)"

58. Listening to the facts spouted by the just-arrived talking antelope, I couldn't believe what the ___ ___ ___

59. Fill in the blank with this word: "Alley ___"

60. Fill in the blank with this word: "___-la-la"

PUZZLE 50

DOWN

1. M.B.A. hopeful's hurdle

2. With 10-Down, bun protectors

3. Wagner's earth goddess

4. Archetype

5. "Please? Please? Please?"

6. Prefix with sclerosis

7. Shipping dept. stamp

8. Be an utter bore?

9. Fabric border

10. Fill in the blank with this word: "___ badge, boy scout's award"

11. Mourning band

12. 1988 Peter Allen musical

15. Fill in the blank with this word: ""With this ring ___ wed""

20. Tetra- plus one

22. Hosp. area

24. Slips

25. If Walls Could Talk...' network

26. Riley's "___ Went Mad"

27. Fill in the blank with this word: ""___ go!""

30. Ronald Reagan's mother

31. Union-busting grp.?

32. Secular

33. Greek gulf or city

34. Texter's 'ciao'

37. Stanford-___ (I.Q. rater)

38. Certain vitamins

40. Not nude

41. Places for moles

42. Fill in the blank with this word: ""___ Tag!""

43. Sue Grafton's '___ for Evidence'

45. Up, in baseball

46. Scold

47. Talk follower

48. Getty Center architect Richard

50. Puccini soprano

51. Fill in the blank with this word: ""___ kleine Nachtmusik""

52. Fill in the blank with this word: "*Ace ___ Stories (old detective pulp magazine)"

53. Unexciting

56. Fill in the blank with this word: "Bad ___, Mich. (seat of Huron County)"

ACROSS

1. Indian butter

5. They cross here

10. Year in which Middle English began, by tradition

13. Zeppo, for one

14. Val d'___, French ski resort

15. Fill in the blank with this word: ""Able was ___...""

16. Presidential ___

17. U.S.A.F. rank

18. "King Lear" or "Hamlet": Abbr.

19. Stomped

21. TV monitors?

23. One-named Belgian cartoonist who created the Smurfs

24. Like some gems

25. When pigs fly!'

28. Tropical fever

29. Lionel to Drew Barrymore

32. Paint base

35. Vampire ___ (fanged fish)

36. What you might hear halting speech in, for short

37. Famously polite Old West stagecoach robber

39. With 4-Down, in relation to

40. the tendency of a body to return to its original shape after it has been stretched or compressed

42. Robt. E. Lee, e.g.

44. L.A.-based petroleum giant

45. Pick up

48. Memory: Prefix

49. 'You don't say?!'

50. They're driven around campsites

54. Twining stem

55. 'Sinatra-___' (1962 album)

57. Stretch ___

58. Wing: Abbr.

59. Surpass

60. Onetime Spanish queen and namesakes

61. The Missing Drink : High ___ rose

62. Sweeties

63. Strange beginning?

PUZZLE 51

ACROSS

1. Fill in the blank with this word: ""What's it all about, ___?""

6. There have been 12 popes with this devout name, the first during the second century, the last from 1939 to 1958

10. Year in Vigilius's papacy

13. Slaves

14. To-do list

15. Years on end

16. 2004 event at which the jinx was broken

18. Neckcloth

19. Fill in the blank with this word: "___-dokey"

20. Where hops are dried

21. Until 1990 it was the capital of West Germany

22. Tear up

26. Forcefully, in music

28. Stub ___

29. Big tournaments for university teams, informally

30. Divisions of a mark

34. Rock's Fleetwood ___

35. Decimal

37. There's ___ in team'

38. Dependent on chance

41. When clocks are changed back from D.S.T. in the fall

43. What's expected

44. Where some jams are made

46. Joe DiMaggio's nickname

50. Fill in the blank with this word: ""Paint the Sky With Stars: The Best of ___" (1997 album)"

51. Typical romance novel love interest

52. West Coast cop squad, for short

55. Fill in the blank with this word: "Debussy's "Air de ___"

56. Assess per person?

59. Fill in the blank with this word: "___ cit."

60. Pierre's girlfriends

61. Fill in the blank with this word: "Andrea ___"

62. In tune

DOWN

1. Fill in the blank with this word: ""A one and ___..."

2. Fill in the blank with this word: "___-see"

3. 1990's sitcom, literally

4. an uncertain region on the east shore of the Adriatic where an ancient Indo-European people once lived

5. Either of two books of the Apocrypha: Abbr.

6. Zoroastrian

7. Question to a consumer watchdog

8. What a lei person might pick?

9. Worrying sound to a balloonist

10. The English translation for the french word: paratonnerre

11. Greene of "Bonanza"

12. Fill in the blank with this word: "___ of itself"

14. Je t'aime : French :: ___ : Spanish

17. Propagates

21. I've ___ had!'

23. Long-running film role

24. Kind of list

25. Robert Burns's

63. Unit of pressure

64. Peak of southeast Nev

"___ Wild Mossy Mountains"

26. Low-cost home loan corp.

27. West Coast sch.

30. Subj. that deals with mixed feelings

31. The English translation for the french word: ingÈrer

32. This meat is also called kid & when curried is a Jamaican specialty

33. Saxophonist Zoot

35. Yielded

36. Quarterbacking locale?

39. Annette Sings ___' (1960 pop album)

40. Water tester

41. Painter/poet Jean ___

42. Round opening

44. Turns state's evidence

45. Something that takes its toll?: Abbr.

46. Mello ___ (soft drink)

47. Particle in electrolysis

48. Town on the Tappan Zee

49. Lorelei, notably

53. Well accessory

54. June 6 1944 is___.

56. Pride : lion :: clowder : ___

57. Wellness grp.

58. William Halsey, e.g.: Abbr.

PUZZLE 52

ACROSS

1. Old Jewish scholars

6. Stored computer images, for short

10. Sci. course

14. Fill in the blank with this word: "Actress Mary Tyler ___"

15. Right turn ___

16. Fill in the blank with this word: "___ contendere"

17. Acetylsalicylic acid and cotton?

20. Was ___ hard on them?'

21. TV's "___-Files"

22. Taliban mullah and others

23. Fill in the blank with this word: "___ rock (radio format)"

25. Fill in the blank with this word: ""___ Simple Man" (#1 Ricky Van Shelton song)"

27. Basic

33. Fill in the blank with this word: "___ good example (shows the proper way)"

34. World Match Play Championship champ a record seven times

35. Rousing cheers

37. It may finish second

38. Burgs

42. Rose ___ rose...'

43. The English translation for the french word: utÈrus

45. Fill in the blank with this word: ""Never Wave at a ___" (1952 military farce)"

46. Fill in the blank with this word: "___ and pains"

48. ... to the river

52. Defeat

53. Fill in the blank with this word: "Dickens heroine ___ Trent"

54. Fill in the blank with this word: ""___ Go Again" (1987 #1 song)"

57. Funnyman Foxx

59. W.W. II vessels

63. Partly tripled, a lyric from a 1964 #1 pop song

66. Speed-skating champ Johann ___ Koss

67. Woe ___!'

68. Fill in the blank with this word: ""This must ___ place""

69. Yep's opposite

70. Teacher's advanced deg.

71. One of the archangels

DOWN

1. Prefix with -vert

2. Yawl or yacht

3. Laddie, in Australia

4. Nipple rings

5. Yak in the pulpit?: Abbr.

6. Israelites' pre-Exodus home

7. Shoe part

8. Willing to talk

9. Prefix with fuel

10. Lively

11. Hurricane-tracking agcy.

12. "Now you're talking!"

13. Unscramble this word: stso

18. Jazz singer ___ James

19. Barnum midget

24. Potato source

26. Where to find the Wienerwald: Abbr.

27. Wood-cutting tool

28. Arrange into new lines

29. Old words from which modern words are derived

30. The English translation for the french word: Delco

31. Grant's first secretary of state ___ Washburne

32. Zero out

36. Window part

39. Seuss's "Horton Hears ___"

40. Chicago's Jane Byrne, once

41. Fill in the blank with this word: "___ Adams, signature on the Declaration of Independence"

44. Vitamin needed for pernicious anemia

47. "Beat the Clock" TV host

49. The muse of history

50. The English translation for the french word: Ècru

51. Work over

54. Personal and direct

55. Craig of the N.B.A.

56. Finish this popular saying: "As you sow so shall you_____."

58. Salinger's 'For ___ - With Love and Squalor'

60. ___ Institute (astronomers' org.)

61. Chevrolet model

62. Illustrator Silverstein

64. When paired with vigor, it signifies exuberance

65. Fill in the blank with this word: ""___ Roi" (Alfred Jarry play)"

PUZZLE 53

ACROSS

1. Finish this popular saying: "If wishes were horses, beggars would_____."

5. Will-___-wisp

9. Wipes clean

14. Quelques-____ (some, in France)

15. Working stiff

16. Perfectly good

17. Expansionist doctrine

20. Jailed, slangily

21. Whip

22. The English translation for the french word: cÈ

23. Fill in the blank with this word: "___ in xylophone"

25. Govt.-issued funds

26. Trick ending

27. Response to the query "Does Ms. Garbo fist-bump?"?

33. Words often before a colon

34. Grand ___ Opry

35. Sacred ___

37. Trick-taking game

38. Town council president, in Canada

41. Go ___ some length

43. Some PX clientele

45. Turner of TV channels

46. Old Fords

47. Book that's a paean to a painter?

51. Tiny bit, in France

53. Shiite center in Iran

54. Slangy turndown

55. Sir ___ McKellen (Gandalf portrayer)

56. Thunderbirds' org.

58. Secured, with "down"

63. Reward of a sort

66. Like many taste tests

67. Stupidity syllables

68. Fill in the blank with this word: "___ Minor"

69. Patty Hearst's name in the S.L.A.

70. Trapped like ___

71. Poet Mark

DOWN

1. Mulling

2. Put ___ appearance

3. Wolves' creations

4. "Scratch that!"

5. Zero personality?

6. French possessive

7. Wilson's "The ___ Baltimore"

8. Range indicator, often

9. Brightly colored pullover garment

10. Last: Abbr.

11. Reduces significantly

12. The English translation for the french word: toner

13. Brilliantly blue

18. Prix ___

19. She, in Italy

24. Unscramble this word: esla

27. Sue Grafton's '___ for Noose'

28. Working in a mess

29. Organized bribery?

30. React violently, in a way

31. Fill in the blank with this word: ""Must-___" (NBC slogan)"

32. Levi's "Christ Stopped at ___"

36. Teen activist org.

39. Way to one's heart

40. She played Leslie Howard's wife in 'Intermezzo'

42. Taoism founder Lao-___

44. Secondary result

48. Fill in the blank with this word: ""I can only ___ much""

49. Orlando's own Mandy Moore

50. Shipping units: Abbr.

51. Titlark

52. Fill in the blank with this word: "___ living"

57. Tea time, perhaps

59. Fill in the blank with this word: "___ Barak, former Israeli P.M."

60. Well, fiddle-dee-dee! He was the fifth emperor of Rome

61. White House's ___ Room

62. June 6 1944 is____.

64. Year in Trajan's reign

65. Shelter grp.

PUZZLE 54

ACROSS

1. Take the top off

6. Time: Ger.

10. Fill in the blank with this word: ""___ Nagila" (song title that means "Let us rejoice")"

14. Pelvis connectors

15. Role for Ingrid

16. Munich's river

17. Year Super Bowl XXXVII was played

18. Nose: Prefix

19. Shoulder muscle, informally

20. Apology #1

23. Water

24. Sci-fi princess

25. Fill in the blank with this word: "___ in apple"

28. Keto-___ tautomerism (organic chemistry topic)

31. Dramatic confession

35. Mop: Var.

37. Expressed surprise

39. Bridge seats

40. See 17-Across

43. Mythical eponym of element #41

44. Harem rooms

45. Spear holder on an opera stage

46. Her feast day is Jul. 11

48. Sylvia ___, whom Sinatra once called the "world's greatest saloon singer"

50. Seabiscuit jockey ___ Pollard

51. The Neverending Story' author

53. Scoot

55. Family financial figure

62. You'll be the death ___!'

63. Golf innovator Callaway and bridge maven Culbertson

64. Sending to the canvas

66. Fill in the blank with this word: "Aleph-___"

67. Warrior whose archenemy was Callisto

68. 1983: "____ and the Cruisers"

69. "Scratch that!"

70. Hard-hitting 1992 hurricane

71. Up on deck

DOWN

1. Mil. branch

2. Fill in the blank with this word: ""What's in a ___?": Juliet"

3. Roman 209

4. Fill in the blank with this word: ""If I Were ___ Man""

5. Nevada tribe

6. Small-circulation publication for fans

7. Jack of "Rio Lobo"

8. Fill in the blank with this word: "Dan ___, former N.B.A. star and coach"

9. Fill in the blank with this word: "___ 101, world's tallest building, 2004-07"

10. Treehouses and such

11. Ralph Vaughan Williams's "___ Symphony"

12. Where many vins come from

13. Vissi d'___' (Puccini aria)

21. Fill in the blank with this word: ""___ a gun!""

22. Word on a soda bottle label

25. Orgs.

26. Watch out for

27. Fill in the blank with this word: ""I thought ___!" ("My feeling exactly!")"

29. Fill in the blank with this word: "___ Valley Conference"

30. Yorkshire city

32. Hit ___ note

33. Dumb bunny

34. Fill in the blank with this word: ""The East ___," song of the Chinese Cultural Revolution"

36. Religious tract?

38. June 6 1944 is___.

41. 59-Acrosses, in Italian

42. Turkish Empire founder

47. Anatomical parts that touch, as eyelids to eyeballs

49. Kind of wrench

52. Moray catcher

54. Fill in the blank with this word: ""Whose ___ these are I think I know": Frost"

55. Traffic ___

56. Fill in the blank with this word: ""The Sum ___" (Russell Crowe movie)"

57. Year in the middle of this century

58. Unit of force

59. Fill in the blank with this word: ""___ Stars," #1 hit for Freddy Martin, 1934"

60. Skirt style

61. Writer Blyton

65. Fill in the blank with this word: ""___ whiz!""

PUZZLE 55

ACROSS

1. John, Paul and John Paul

6. Harry Hershfield comic "___ the Agent"

10. It's in the back row, right of center

14. Student with the motto 'Fiat Lux,' informally

15. Suffix in nuclear physics

16. Put ___ words

17. ANTE

20. 1998 Wimbledon winner Novotna

21. More gung-ho

22. Work without ___

23. Balkan land: Abbr.

24. Dried seaweed popular in Japanese cuisine

28. Where Douglas MacArthur returned, famously

30. Wishful thinker of story

32. Snug snack item?

35. The English translation for the french word: ecstasy

36. Caution against neglect

40. Test result, at times: Abbr

41. Fill in the blank with this word: ""But thy ___ summer shall not fade": Shak."

42. Wrong

45. Means

49. The English translation for the french word: recruter

50. Shakespeare's ___ of Salisbury

52. Kangaroo ___

53. Physicist who pioneered electromagnetism

56. The Green Hornet's valet

57. Russia, China and France are in it

61. Politico Hutchinson and others

62. Fill in the blank with this word: "Big ___ (nickname of baseball's Ortiz)"

63. Psychiatrist/author R. D. ___

64. Yawn

65. Fill in the blank with this word: "___ lot (gorged oneself)"

66. Fill in the blank with this word: "Cop ___"

DOWN

1. Land south of Kashmir

2. Setting of 1984

3. Unscramble this word: npealt

4. I could ___ horse!'

5. Winter weather, in Edinburgh

6. Like some chambers

7. Fort ___, N.C.

8. Actress Skye

9. M.I.T. grad: Abbr.

10. Lamebrain, in slang

11. Log cabin material, maybe

12. BBC rival

13. Finish this popular saying: "Waste not want_____."

18. Undoes a breakup

19. Reggae's ___-Mouse

23. Adriatic seaport

25. Year Cort

26. verb secure with a bitt

27. The Tar Heels: Abbr.

29. Kwik-E-Mart clerk on "The Simpsons"

30. "Now you're talking!"

31. Virgin of the Rocks' painter

33. Written reminder

34. Test in coll., perhaps

36. Tucson sch.

37. Org. for the Denver Gold and Chicago Blitz

38. The English translation for the french word: saul

39. Taina who was one of Les Girls, 1957

40. Fill in the blank with this word: "___ favor"

43. Top of a closet?

44. Yesterday, in the Yucatán

46. RNA constituent

47. Rough, loosely woven fabric

48. Her feast day is Jul. 11

50. Classic British Jaguar

51. Fungal spore sacs

54. TV star who shills for Electrolux

55. New York's ___ Island

56. Sack starter

57. Wilt

58. Fill in the blank with this word: "Conductor ___-Pekka Salonen"

59. With 13-Down, they go off with a bang

60. Suffix with form

PUZZLE 56

```
 1  2  3  4  █  5  6  7  8  █  9 10 11 12 13
14          █ 15          █ 16
17          █ 18          19
20          21 █    22          █ 23
24             25 26          27       █
█  28             29          █ 30 31 32 33
34          █ 35 36          37 38
39          40             41
42             █ 43          █ 44
45          █    46          47 48       █
█    49 50 51          52             53
54 55 56    █ 57          █ 58
59          60          61 62    █ 63
64          █    65          █ 66
67          █    68          69
```

ACROSS

1. Yearly loan figs.

5. Fill in the blank with this word: ""___, Our Help in Ages Past" (hymn)"

9. Weenie

14. Oscar-nominated actress for "Leaving Las Vegas"

15. Of port, starboard, fore or aft, the one usually yelled by a golfer

16. Scarlett ___ of "Gone With the Wind"

17. The English translation for the french word: mime

18. Sprouts

20. "The Tempest" king

22. Fill in the blank with this word: ""___ to Billie Joe" (1967 #1 hit)"

23. Sue Grafton's "___ for Alibi"

24. "Date film" classic of 1987

28. Where a queen may be crowned

29. See 20-Across

30. Nearly

34. Har-___ (tennis surface)

35. Fill in the blank with this word: "___ off (switch choice)"

37. Response to "Are you awake?"

39. Fact about unladylike habits?

42. Mother of Xerxes I

43. With 21-Across, like many rivers in winter

44. Y. A. Tittle scores

45. View from a beach house

46. TV announcer Hall

47. Japon's place

49. See 23-Across

54. Paul Scott series "The ___ Quartet"

57. TWA rival

58. Lead-in to a questionable opinion

59. Government resister standing ready

63. How to ___ knot (Boy Scout's lesson)

64. This German-language writer's novels were all published after his 1924 death, including "The Trial"

65. Island in French Polynesia

66. Fill in the blank with this word: "Author ___ S. Connell"

67. Actually

68. Fill in the blank with this word: ""To thine own ___ be true""

69. D.C. group

DOWN

1. Equally irate

2. "Portnoy's Complaint" author

3. Gossipy bartender's choices?

4. Taken care of

5. Fill in the blank with this word: "Attila, the Scourge ___"

6. Stuff

7. Scottish Peace Nobelist John Boyd ___

8. Person who likes the blues?

9. Stanley who co-directed "Singin' in the Rain"

10. Fill in the blank with this word: ""___ hoppen?""

11. I could ___ horse!'

12. Riley's "___ Went Mad"

13. Word with blue or sea

19. Uganda's ___ Amin

21. Arthur who wrote "The Symbolist Movement in Literature"

25. Testify

26. Sea nymphs of myth

27. Fed. lending agency

31. Enterprise

32. Aesthetically pleasing ratio of antiquity

33. Rtes.

34. Fill in the blank with this word: ""___ wondrous pitiful": "Othello""

36. Wiretapping grp.

37. Wrath

38. Title family name on TV

40. Dusseldorf donkey

41. Fill in the blank with this word: ""___ Live," 1992 multiplatinum album"

46. Unagi restaurant suppliers

48. Treats with 26-Down

50. Fill in the blank with this word: "___ sea (cruising)"

51. Not the usual spelling: Abbr.

52. Fill in the blank with this word: "___ Olay"

53. Has kids

54. Turkish brandy

55. Josephine Tey investigator ___ Grant

56. One of the acting Bridges

60. QB Manning

61. Worker in a garden

62. Suffix with ether

PUZZLE 57

ACROSS

1. a name for the God of the Old Testament as transliterated from the Hebrew consonants YHVH

7. Sticker letters

11. Turn signal dirs.

14. Like some salad dressings

15. Runs a tab

16. Finish this popular saying: "You are what you_____."

17. Unlock

18. "Qu

19. Fill in the blank with this word: "___ Pictures (old studio)"

20. #1 hit of 1956

23. Rap sessions?

26. Suggest

27. Pupil's place

28. Reply facilitator: Abbr.

30. Frees from

31. Young salamanders

33. Like a suit with a vest

35. Spirit of an evil evil spirit?

40. Fill in the blank with this word: "El ___"

41. Tiny time unit: Abbr.

43. States with authority

46. Startled cry

48. Fill in the blank with this word: "1968 hit "Harper Valley ___""

49. Fill in the blank with this word: ""You're ___ and don't even know it""

50. Like some politics

53. Some story collectors

57. Year that Eric the Red was born, traditionally

58. Film director Nicolas

59. Some war plans

63. Fill in the blank with this word: ""Some ___ meat and canna eat": Burns"

64. Ring figure

65. Prefix with bacteria

66. See 38-Across

67. Had an unquiet sleep

68. They worship Jah

DOWN

1. Tenth letter of the Hebrew alphabet

2. Work without ___

3. Sot's sound

4. Long-crested bird

5. The English translation for the french word: douer

6. Villains in 'The Lion King'

7. Water under the bridge

8. Wield a mop

9. Cut down on

10. Ones charging reading fees?

11. You've seen them before

12. Start liking

13. Minnesota college

21. This first name of Miss Marple's creator comes from a word that means "good"

22. Continue the journey

23. Somme sight

24. One of the acting Bridges

25. Vertical departure, acronymically

29. Lyric poem

30. Theologian's subj.

32. Washes with detergent

34. African fly

36. Making daguerreotypes, and other things

37. Miracle-___

38. Org. whose workers may be left carrying the bag

39. Fill in the blank with this word: ""I ___ Song Go Out of My Heart""

42. No-___-do

43. I Really Like Him' singer in 'Man of La Mancha'

44. Transitional primate

45. Cried out in pain

47. Fill in the blank with this word: "___ roll"

51. Spanish queen

52. Need for the winner of a Wimbledon men's match

54. Tough

55. Fill in the blank with this word: "___ Aarnio, innovative furniture designer"

56. Haing S. ___ (Oscar winner for 'The Killing Fields')

60. Fill in the blank with this word: ""Did you ___ that?""

61. Victorian ___

62. Fill in the blank with this word: "___ pad"

PUZZLE 58

ACROSS

1. Fill in the blank with this word: ""Mi casa ___ casa""

5. Trade

9. Yawn

12. Quaint contraction

14. Fill in the blank with this word: "___-miss"

16. Japanese prime minister Taro ___

17. Reprimands

19. When clocks are set ahead: Abbr.

20. Unreal

21. water causes the peg to swell and hold the timbers fast

23. With the intent

25. Sleep like ___

26. Twaddle

30. Vulgarity

33. Fill in the blank with this word: ""So ___ me!""

34. Betelgeuse, for one

36. The P.L.O.'s Arafat

37. Like squads in arena football

39. Formula ___

40. Sales slip: Abbr.

41. What debaters debate

43. Fill in the blank with this word: "___ poker (bar game)"

46. Ursine : bear :: pithecan : ___

47. Spinachlike potherb

49. Quick movements

51. Nose: Prefix

52. Fill in the blank with this word: "___ Millions (multistate lottery)"

53. Norm's last name, on "Cheers"

57. Meeting of leaders

61. Fill in the blank with this word: "___ Friday's"

62. Contract bridge tactics

64. Popular cable channel

65. Cousin of the needlefish

66. That's an order!

67. Fill in the blank with this word: ""It ___; be not afraid" (words of Jesus): 2 wds."

DOWN

68. Revenuer

69. Terhune's "___ Dog"

1. Villa d'___

2. Fill in the blank with this word: "___-Pei (dog)"

3. Fill in the blank with this word: "___ Fifth Avenue"

4. Trepidation

5. Fill in the blank with this word: ""___ 'nuff!""

6. 1984-88 skating gold medalist

7. essential oil or perfume obtained from flowers

8. Stand in for

9. Its currency unit is the ariary

10. They, in Italy

11. Wilson's "The ___ Baltimore"

13. Violent weather, informally

15. Singer with the 1994 #1 hit "Bump N' Grind"

18. Russian autocrats: Var.

22. Lack of musical talent

24. Yuccalike plant

26. Of ___ (availing)

27. Surgeon's target

28. Groups battling big government

29. Patty Hearst's name in the S.L.A.

31. Nothin'

32. The sculptures "Rigoletto" and "La Tosca," e.g.

35. This synonym for "kingdom" comes from the Latin for "regal"

38. The English translation for the french word: niche

42. Trilled calls

44. Like non-oyster months

45. Made a smooth transition

48. Put on the throne

50. Cut capers

53. shaper of the world

54. William ___, Alaska's first 72-Down

55. The English translation for the french word: ovule

56. Tiber tributary whose name means "black"

58. Fill in the blank with this word: ""Mississippi Masala" director ___ Nair"

59. What ___ for Love' ('A Chorus Line' song)

60. Puccini soprano

63. Rembrandt van ___

PUZZLE 59

ACROSS

1. Fill in the blank with this word: ""...partridge in ___ tree""

6. Wedge-shaped inlet

9. West Coast cop squad, for short

13. French income

14. You may part with it

15. Fill in the blank with this word: "Europe's Gorge of the ___"

16. "Labyrinth" star, 1986

18. Prized game fish

19. Deception

20. Spore cases

21. Western N.C.A.A. powerhouse

24. Use in great excess

25. Victimizes, with 'on'

26. Devil, to Muslims

28. Wing-shaped

29. Surface again, as a road

30. Star Trek' lieutenant

33. Want-ad letters

36. Start of an Oscar Wilde quote

39. Westminster Show org.

40. Seacrest's 'American Top 40' predecessor

41. "Don't even bother"

42. Fill in the blank with this word: "Banda ___ (2004 tsunami site)"

44. Loser

46. Jason who directed 2011's "Arthur"

48. Fill in the blank with this word: ""You're ___, ya know that?": Archie Bunker"

49. Yankee insignias

50. Poet ___ Van Duyn

51. Vets

54. Hydrocarbon suffixes

55. Frank request

59. Yorkshire river

60. Fill in the blank with this word: "___ out a living (scraped by)"

61. Takes in

62. Russia's Itar-___ news agency

63. Fill in the blank with this word: "___ el Amarna, Egypt"

64. [See title, and proceed]

DOWN

1. Suffix with tank

2. Fill in the blank with this word: "___ green"

3. Stationer's item: Abbr.

4. Suffix with symptom

5. Painful prod

6. Mr. Bean portrayer Atkinson

7. What a rare mood ___'

8. Have ___ in one's bonnet

9. Russian port, formerly Kuibyshev

10. Red flag for the I.R.S.

11. Dear

12. St.-___, capital of R

14. Milk sources

17. Fill in the blank with this word: "___ fond farewell to"

21. Std. on food labels

22. Valentino title role, with "the"

23. Hospital imaging devices

25. Frank

27. The English translation for the french word: inversion de contrŬle

28. Fill in the blank with this word: ""If the ___ is concealed, it succeeds": Ovid"

31. Fill in the blank with this word: ""___ Haw""

32. (of unknown regions) not yet surveyed or investigated

34. Key's opener?

35. Women's dress sizes

37. Yup's alternative

38. Michigan's ___ Canals

43. Winds up

45. Way to get to N.Y.C.

46. Wing or breast

47. Where Ephesus was

48. Fill in the blank with this word: "Challenge to ___"

51. Kind of romance between actors

52. While you're enjoying the mountain scenery you might spot a tarn, a small one of these left by a glacier

53. Women of Andaluc

56. Fill in the blank with this word: "___ Onassis, Jackie Kennedy's #2"

57. German link

58. Taoism founder Lao-___

PUZZLE 60

(crossword grid)

ACROSS

1. Margaret Mead's "Coming of Age in ___"

6. Ways: Abbr.

10. Work at the docks

14. Lymphocyte found in marrow

15. Soprano Berger

16. Fill in the blank with this word: "___ off (switch choice)"

17. Club

18. This can be a careless mistake or a foolish person who might make one

19. Powerful kind of engine

20. Disingenuous sorrow

23. Paint base

24. Least wild

25. Overstep a boundary

31. Bad losers

32. Like some internships, in length

33. Fill in the blank with this word: "___ Jones"

36. Tempura ___ (Japanese dish)

37. Levels

38. The kissing disease'

39. Zoologist's foot

40. The Wizard of ___ Park

41. UnitedHealth rival

42. Herring on a fishhook?

44. Win ___

47. Sounds of doubt

48. End of the headline

54. Gray ___

55. Take ___ face value

56. Rose-red dye

58. Kipling's "___ we forget!"

59. Fill in the blank with this word: "De ___"

60. "Little" girl of old comics

61. Fill in the blank with this word: "___-majestÈ"

62. Those, to Tom

63. #1 song

DOWN

1. Year that Dionysius of Halicarnassus is believed to have died

2. Fill in the blank with this word: ""___ Live," 1992 multiplatinum album"

3. "You've Got Mail" actress

4. Half an old vaudeville duo

5. Not involving check or credit

6. Officially listed: Abbr.

7. U.S.S. Enterprise counselor

8. Keto-___ tautomerism (organic chemistry topic)

9. Two-point plays in football

10. 1956 Elvis hit that went to #2

11. Temple architectural features

12. They aren't just talkers

13. Fill in the blank with this word: "007 foe ___ Blofeld"

21. Mel who was #4 at the Polo Grounds

22. Vehicle with a rotating top

25. Burger topper

26. Went on

27. Top prizes at the Juegos Ol

28. When doubled, a former National Zoo panda

29. Fill in the blank with this word: "1940's-50's All-Star pitcher ___ Blackwell"

30. Fill in the blank with this word: ""Vive ___!""

33. The lady ___ protest too much': Shak

34. Tony-nominated choreographer White

35. The English translation for the french word: guËde

37. Largest of the Canary Islands

38. Where things are bolted down on base

40. Tangled and interwoven

41. Tonto portrayer, briefly

42. The English translation for the french word: dilater

43. Wheeler Peak locale: Abbr.

44. Headed for ___ (in imminent trouble)

45. Thick soup

46. Like non-oyster months

49. Truncation indications: Abbr.

50. Roman statesman ___ the Elder

51. Western Indians

52. Fill in the blank with this word: "___ noche (tonight, in Tijuana)"

53. Telegraph clicks

57. Yup's alternative

PUZZLE 61

ACROSS

1. What to wear on dress-down day

8. 1940's-60's world leader

13. Inability to smell

14. Any installation designed to accommodate patrons in their automobiles

16. Send documents or materials to appropriate destinations

17. Makes airtight again

18. Arrays

19. Adaptable

20. Comparative word

21. Gabriel, for one

22. "___ Town Too" (1981 hit)

23. M√°laga man

25. Deteriorate

27. "The Three Faces of ___"

28. A feeling of profound respect

31. "___ what?"

32. Really regret

33. Infomercials, e.g.

36. Stereo type?

40. Backstabber

41. Baptism, for one

42. Digress

44. "The Catcher in the ___"

45. Order to attack, with "on"

46. Arid

47. Brewer's product

49. Czech composer

52. An event that departs from expectations

53. Freshens, in a way

54. Recluse

55. Most foul

56. Give an interpretation or explanation to

57. Subscription cards, e.g.

DOWN

1. Fifths of a gram

2. Bar order

3. A long cassock with buttons down the front

4. Six-time portrayer of Poirot

5. Gulf V.I.P.

6. Ancestry

7. Declines

8. Month after Adar

9. Times to call, in classifieds

10. Unenlightened one

11. Sculpture consisting of shapes carved on a surface

12. Turn left, often

14. "No jeans" may be part of one

15. Someone who relates or narrates

24. 1990 World Series champs

25. Make a copy of

26. Dark

29. Accustom anew

30. Adjusts, as a clock

33. Someone who arrives (or has arrived)

34. Working mom's aid

35. Guided

37. On the line

38. One buying a round

39. Heartfelt

43. Brewers' supplies

45. Composed

48. Kind of store

49. A dress worn primarily by Hindu women

50. Aim

51. Coastal raptors

PUZZLE 62

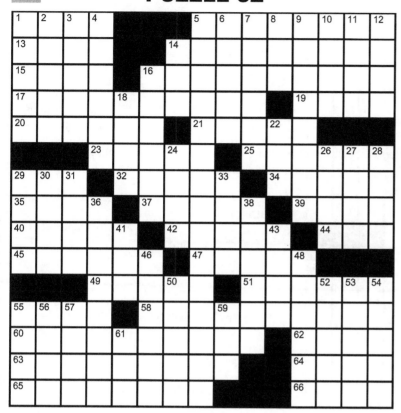

ACROSS

1. Quarters

5. Large-oared craft on a ship

13. Outfielder Mondesi

14. Smoke with straight sides

15. Fill in the blank with this word: "___ arch"

16. Is nosy

17. Rearranges the lettuce?

19. Corp. money managers

20. Alphabet sextet

21. Body of water in a volcanic crater, for one

23. The English translation for the french word: synode

25. Oust from office

29. Fill in the blank with this word: "i___ boom bah!î"

32. With Altair and Vega, it forms the Summer Triangle

34. Money for Amer. allies

35. Zebulun's mother, in the Bible

37. Arrange into new lines

39. She, in Italy

40. Options at a gym

42. Wicked Game' singer Chris

44. Fill in the blank with this word: ""___ will be done""

45. Vowel sound in "puzzle"

47. Gradually quickening, in mus.

49. Fond ___, Wis.

51. Tricky fellows

55. Russian gold medalist ___ Kulik

58. Jungle gym's place

60. Items filling a star's mailbox

62. Prized game fish

63. Personal ad info

64. Zeno of ___

65. Some fortified wines

66. Twinkies or cookies, e.g.

DOWN

1. Erich who wrote "The Art of Loving"

2. Include as an extra

3. Ingredient in a Spanish omelet

4. Visits dreamland

5. Scented pouches

6. Jermaine ___, six-time N.B.A. All-Star

7. Your line of fate is quite deep, indicating success investing with tech stocks, like Adobe & Oracle, on this exchange

8. Ensured: Abbr.

9. Cradlesong

10. Two-time Swedish prime minister Palme

11. Fill in the blank with this word: "___ Snaps (dog treats)"

12. Russia's Itar-___ news agency

14. Oom-___ (polka rhythm)

16. Wage ___

18. Wallpaper meas.

22. Nickname of 1954 home run leader Ted

24. The ___ Love' (R.E.M. hit)

26. Yesteryear

27. They take the bait

28. Four hours on the job, perhaps

29. Stallone and others

30. Fill in the blank with this word: "Architect ___ Ming Pei"

31. Fill in the blank with this word: "___ palm"

33. U.K. carrier, once

36. Helm cry after "Ready about"

38. Flakes

41. Fill in the blank with this word: "Disco ___ (character on "The Simpsons")"

43. Fill in the blank with this word: "___-Ration (dog food)"

46. Woodstock, N.Y., county

48. They're no experts

50. Fill in the blank with this word: "___ part (role-plays)"

52. Actor Raf of "The Italian Job," 1969

53. Ohio natives

54. -

55. Fill in the blank with this word: ""___ not back in an hour...""

56. Pop singer ___ Del Rey

57. Having depth

59. Surgery sites, for short

61. Verdi's "___ tu"

PUZZLE 63

ACROSS

1. Greeter at the door

11. Songwriter Novello

15. Energize

16. Fill in the blank with this word: ""___ #1!""

17. There are six of these in the middle of 17- and 56-Across and 11- and 25-Down

18. Yard sale tag

19. Something to chew on

20. That ole boy's

21. Its slogan was once "More bounce to the ounce"

22. You'll want to munch on petha & gazak, signature sweets of this Taj Mahal city

24. Nautical direction

27. Pizzeria ___ (fast-food chain)

28. "Quiet!"

32. Sports org. that publishes DEUCE magazine

33. Whse. unit

34. TV remote, e.g.

36. Fill in the blank with this word: "___ in Charlie"

37. It may be pulled

41. Mother ___

42. Deep fissure

43. Command level: Abbr.

44. Kellogg's Cracklin' ___ Bran

45. Your line of fate is quite deep, indicating success investing with tech stocks, like Adobe & Oracle, on this exchange

48. Fill in the blank with this word: "___ Aquarids (May meteor shower)"

49. Fill in the blank with this word: ""___ can you see ...?""

51. Pas ___ (dance solo)

53. Turban & this flower name share the same Turkish roots

55. ___ The Magazine (bimonthly with 35+ million readers)

59. Sports org.

60. Uzbekistan's ___ Sea

61. "Whipped Cream & Other Delights" frontman

64. Nesters

65. Wipe out

66. Young lady of Sp.

67. Time for playoffs

DOWN

1. So-called "white magic"

2. Uncle!'

3. French Bluebeard

4. What secondary recipients of e-mails get

5. Texter's 'Alternatively ...'

6. Subcompact

7. Spacewalks, to NASA

8. This unusual food falls from heaven for 40 years starting in Exodus 16

9. Part of an 800 collect call number

10. French possessive

11. Fill in the blank with this word: ""___ a Teen-age Werewolf""

12. He demonstrated that what Columbus had discovered was not 6-Down

13. Fill in the blank with this word: "___ shorthair (cat breed)"

14. Unscramble this word: oseserpn

21. Wow

23. African fox

25. Where Douglas MacArthur returned, famously

26. Mother of Xerxes I

29. Poultry place

30. Political extremists

31. Una ___ (old coin words)

35. Year in Septimius Severus's reign

37. They're spotted in Africa

38. University teacher

39. Medical inspiration?

40. Revenuers

46. Virgilian hero

47. Book size, in printing

50. Well-knit tales

52. Grow dark

54. Role for Ingrid

56. This won't hurt ___!'

57. Stadium sounds

58. Graceful bend

61. World War II Air Force commander ___ Arnold

62. Rock's Brian ___

63. Mao's mil. force

PUZZLE 64

ACROSS

1. Losers

9. They have many cuts, typically

15. Study of Louis Pasteur

16. Mount ___

17. Medical inspiration?

18. Daughter of Poseidon who was the ancestor of a prophetic clan

19. Umberto ___, author of 'The Name of the Rose'

20. They may be good for singles

22. In ___ (where found)

24. In ___ of (replacing)

25. Writer Ephron

26. Lawyers: Abbr.

28. Politically incorrect endings

30. Fill in the blank with this word: "___ Park, home for the Pittsburgh Pirates"

32. Stockpiler

34. Tiny bit

35. Top dogs

38. Words of denial

40. Whim

41. Shark's target, at times

45. The English translation for the french word: CGI

46. Tissue: Prefix

47. Sporty Spice's other nickname

50. Ancient Semitic fertility goddess

52. Town on the Thames

54. Sound system brand

56. Hunters' decoys

59. Usher's offering

60. Salon tints

61. Gave missiles to

63. Pushkin's 'Eugene ___'

64. Show concern for, in a way

65. TD Waterhouse online competitor

66. Powerful speakers

DOWN

1. Fill in the blank with this word: ""___ Irish Rose""

2. Italian automaker since 1906

3. Controversial 1980s-'90s baseball team owner

4. Whit

5. Former Connecticut governor Jodi

6. Thin as ___

7. Chuckleheads

8. Shiny fabrics

9. Swear to

10. Writers Meyer and Ira

11. The Real Housewives' network

12. Support

13. Global agricultural company

14. 1983 Indy winner Tom

21. Wind-swept hill

23. Wasatch County resident

27. Fill in the blank with this word: ""An' singin there, an' dancin here, / Wi' great and ___": Burns"

29. Nobelist poet ___ Axel Karlfeldt

31. This, to Th

33. Spore cases

34. Some church music

35. Mandela's org.

36. any connection or unifying bond

37. Aunt in "Arsenic and Old Lace"

39. Fill in the blank with this word: ""Roll ___ bones!""

42. Tiny fraction of a min.

43. Some Ontario natives

44. Most imperturbable

46. One of two sides of a story?

48. Wilderness home

49. Your professional life, hopefully a distinguished one

51. The English translation for the french word: tonga

53. Body of water in a volcanic crater, for one

55. Starfleet V.I.P.'s: Abbr.

57. Old laborer

58. Fill in the blank with this word: "Attention ___"

60. Weeder's tool

62. Thomas Moore poem "___ in the Stilly Night"

PUZZLE 65

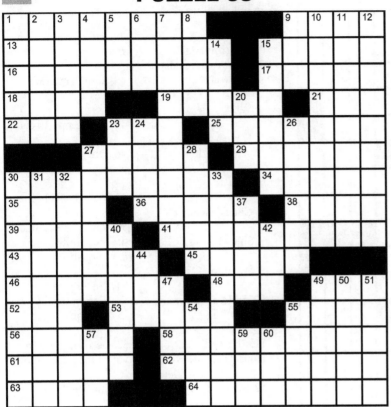

DOWN

1. Wall supports

2. Odd jobs

3. Words spelled out in currants on a Wonderland cake

4. Prized game fish

5. Washington State's Sea-___ Airport

6. Rock-___, classic jukebox

7. a British lawyer who speaks in the higher courts of law

8. Veteran journalist ___ Abel

9. Where Oskar Schindler is buried: Abbr.

10. Heavy metal venue?

11. Town south of San Bernardino

12. From this moment on

14. Tiny battery

15. Purity units

20. The Beatles' "___ Mine"

23. Tea time, maybe

24. Golfers' delights

26. Squeal on

27. Irish runner Coghlan

28. Goddess of agriculture

30. Bartender?

31. It's not too bright

32. Cash cache, often

33. 1960's TV war

drama, with "The" [2008]

37. Glazed, waxy fabric finish

40. Single-minded theorizer

42. The Pres., militarily speaking

44. Tic-tac-toe side

47. Fill in the blank with this word: ""___ Man," Emilio Estevez movie"

49. Fill in the blank with this word: "___ Coyote"

50. Deprives of judgment

51. Thrown for ___

54. His ___ (cribbage jack)

55. The English translation for the french word: Joîl

57. What a constant hand-washer probably has, for short

59. Three-way joint

60. Boatload

ACROSS

1. Appear

9. Venezuela's ___ Margarita

13. Skipper's run

15. Capital of Jamaica [black]

16. Hives, medically

17. Fill in the blank with this word: ""You ___ Lucky Star" (1935 #1 hit)"

18. The English translation for the french word: putain

19. Show, but not premiere

21. Suffix with ether

22. Wind dir.

23. Without a Trace' org.

25. City in a George Strait country hit

27. Protestant denom.

29. Brightest star in Draco

30. a metric unit of length equal to ten meters

34. What angry bees do

35. Haphazard

36. Polar irregularity

38. Ye ___ Shoppe

39. Woolf's "___ of One's Own"

41. Netlike

43. U.S. bullion site

45. Sail spar

46. Turpentine, e.g.

48. Truman's nuclear agcy.

49. Tate and Bowe were once champions of it: Abbr.

52. Sault ___ Marie

53. Fill in the blank with this word: ""___ a Letter to My Love" (Simone Signoret film)"

55. Well-intentioned girl of rhyme?

56. Objects carrying magic spells

58. What an investor builds

61. Put on ___ (pretend)

62. Wind player's purchases

63. Unlikely candidate for prom king

64. Like tank tops

PUZZLE 66

ACROSS

1. Swallowing of food, e.g.

10. Used to entangle a cow's legs, gauchos make good use of this weapon of strong cords with weighted ends

14. Places for mending

15. The Bahamas' Great ___ Island

17. Plot

18. Tried to keep one's seat

19. Philosopher David

20. Northwest Terr. native

21. Elegantly groomed

22. Silver ___

23. "Romanian Rhapsodies" composer

25. Fill in the blank with this word: "___ Arnold's Balsam (old patent medicine)"

26. French possessive

27. W. C. Fields film "___ a Gift"

28. Patio grill

30. The Unsers of Indy

31. One in a balloon

32. Smear with wax, old-style

34. Fill in the blank with this word: "___ 1 (Me.-to-Fla. highway)"

35. Some hallucinogens, for short

36. Got a slice of

38. The Carolinas' ___ Dee River

39. Where the Raptors play

40. Fill in the blank with this word: ""Chances ___," 1957 #1 hit"

41. Makes me want seconds!'

44. Got ___ deal (was rooked)

45. East African, old-style

47. XXX counterpart

48. Put safely to bed, as a child

50. Vatican vestment

51. Police dept. employee

52. Wedding helper

53. When arguments begin?

55. Words said upon departure

56. Given life, perhaps

57. Sound of a leak

58. They snip and clip

DOWN

1. Doesn't work anymore, informally

2. Unscramble this word: tnuaer

3. Garden decorations

4. Yeats's land

5. What a mess!

6. Variety show showing

7. Van Gogh work

8. They're heard in a pen

9. Scottish refusal

10. Prefix with -meter

11. Homage

12. Fund designation

13. Ornate leaf, or a design patterned after it

16. Wonderful musical output?

21. Fill in the blank with this word: ""Bon ___ ""

23. Upscale women's clothes: ____ Fisher

24. Tony Hillerman detective Jim

29. Tots' hiding game

30. Certain navigational aids

31. When some stores open

32. Glee clubs

33. Symptoms of otitis

34. Read the ___ act

36. Unscramble this word: tausst

37. Call in the game Battleship

38. Kind of court

40. Bedridden, say

41. State of Grace

42. Zero on the screen

43. Zippy rides

46. Ran fast, to Brits

49. Some nest eggs

51. Fill in the blank with this word: ""___ idea!""

53. Taoism founder Lao-___

54. Fill in the blank with this word: ""___ Me Call You Sweetheart""

PUZZLE 67

ACROSS

1. Tony winner Neuwirth

5. "King Lear" or "Hamlet": Abbr.

9. Pakistan's chief river

14. Now ___ me down to sleep'

15. Actress Skye

16. Fill in the blank with this word: "___ in sight"

17. Plight of an overcrowded orchestra?

20. Vocal quality

21. This ___ outrage!'

22. Renault, e.g.: Abbr.

23. WWII sphere

25. Verdi's "___ tu"

26. Zine staff

27. Clotheshorses

33. TV star who shills for Electrolux

34. Fill in the blank with this word: ""___ whiz!""

35. Harsh cry

37. John Irving's "A Prayer for ___ Meany"

38. "Hamlet" courtier

41. Fill in the blank with this word: ""The Worst ___ in London" ("Sweeney Todd" song)"

43. Fill in the blank with this word: ""___ Live," 1992 multiplatinum album"

45. Siglo de ___ (epoch of Cervantes)

46. Mechanic's ___

47. End of the quote

51. Org. for mom-and-pop stores

53. Caveat on a party invitation: Abbr.

54. Intimate

55. Switzerland's ___ Leman

56. 2nd qtr. starter

58. Wipes out

63. Fill in the blank with this word: ""The reporter heard the New York ___ ___ his coach""

66. Once-popular Olds

67. City near the Caspian Sea

68. Gold-medal Olympic swimmer ___ Torres

69. Joy Adamson's

"Forever Free: ___ Pride"

70. Snatches

71. Fill in the blank with this word: ""The ___ the limit!""

DOWN

1. verb secure with a bitt

2. Textbook market shorthand

3. Malt liquor yeast

4. Look at

5. See you later!'

6. Shad ___

7. Fill in the blank with this word: "Decem ___ (Latin decade)"

8. Theodor ___ (Dr. Seuss)

9. Stupid joke

10. Finish this popular saying: "Waste not want ___."

11. Poker variation

12. Not mentioned, as a bridge suit

13. Dairy Queen offerings

18. Riley's "___ Went Mad"

19. The English translation for the french word: Nara

24. Fill in the blank with this word: "___ place"

27. To and ___

28. Sony subsidiary

29. Doc wears them

30. Coffee additive

31. Victim of a 1955 coup

32. To Helen' writer, in footnotes

36. Sushi bar cupful

39. High-performance Camaro ___-Z

40. Sounds from the lea

42. 1983 Indy winner Tom

44. Where suits are put on

48. Fill in the blank with this word: ""___ Blue""

49. Like Samsung Corporation

50. Vega's constellation

51. Unscramble this word: islec

52. Stew flavorer

57. Fill in the blank with this word: "___ la Douce"

59. Fill in the blank with this word: ""No ifs, ___ ...""

60. Wyo. neighbor

61. Weird: Var.

62. Women of Andaluc

64. The Era of ___ (1964-74 Notre Dame football)

65. Watership down?

PUZZLE 68

ACROSS

1. Short hair, to Burns

6. Warm-up for the college-bound

10. Naut. law enforcers

14. Stock phrase

15. Feather, to Fernando

16. Neighborhood west of the Bowery

17. Assail rioters dressed in gray, palindromically

20. Poe's "The Murders in the ___ Morgue"

21. Peer group?

22. Grafton's '___ for Malice'

23. One earning rewards

27. Deferential

28. Fill in the blank with this word: "___ Digital Shorts (late-night comic bits)"

29. Je ne ___ quoi

30. Meows : cats :: ___ : dogs

32. Years on end

34. Young hijos

38. In front of a hydrant, say

42. Fill in the blank with this word: "___ a fox"

43. Popular 1980s arcade game based on simple geometry

44. Follows

45. Old Fords

48. Fill in the blank with this word: ""Thanks a ___!""

50. Slowing, in mus.

51. Dairy frivolity?

55. Three ___ match

56. Workplace fairness agcy

57. Taoism founder Lao-___

58. Judges the crying of comic Johnson?

64. Not worth ___ cent

65. Yawning

66. Religious leader ___ Muhammad

67. Vega's constellation

68. Young lady of Sp.

69. Golf's ___ Cup

DOWN

1. W. S. Gilbert's "The ___ Ballads"

2. Fill in the blank with this word: ""Am ___ believe Ö?""

3. Fill in the blank with this word: "45 ___"

4. This cavalry weapon was inspired by the Turkish scimitar

5. Unhip person

6. Stamp not reqd.

7. Struck down

8. Unwelcome financial exams

9. Point ___ (southernmost point in continental Europe)

10. Free

11. W.W. I battle site

12. We'll all sing your praises when you spell...

13. Oodles

18. Got back together

19. Bedridden, say

23. Wild

24. Put in hot oil again

25. Wells's oppressed race

26. Losing ___ Washington

27. Google ___

31. Viking poet

33. Place to hang a ring

35. Their ranks don't include DHs

36. Golden ___

37. Way: Abbr.

39. Lauder and namesakes

40. Sketch

41. Lives

46. Rastafarian's do, for short

47. Dormitory annoyance

49. Richard ___, director of "Help!" and "A Hard Day's Night"

51. Like certain math operations

52. Post office gizmo

53. Stravinsky's "___ for Wind Instruments"

54. Fill in the blank with this word: "1992 Pulitzer historian Mark ___ Jr."

55. Word-of-mouth

59. Soprano Christiane ___-Pierre

60. Water tester: Abbr.

61. Grant-in-___

62. Union activist Norma ___ Webster

63. Nasdaq unit: Abbr.

PUZZLE 69

ACROSS

1. Mountain ___ (Pepsi products)

5. ___ Church, country singer with the #1 hits 'Drink in My Hand' and 'Springsteen'

9. Stock market figures

14. One way to race

15. Fill in the blank with this word: "Crystal ___"

16. Fill in the blank with this word: "___ part (role-plays)"

17. Flip response?

18. Biblical dry measure: Var.

19. Fill in the blank with this word: ""___ say...""

20. L/L Bean?

23. Unrealized 60's Boeing project

24. Fill in the blank with this word: "___ Khan"

25. Like some races and hopes

26. Old what's-___-name

29. Animal lab technician's work?

33. What "two" meant, historically

34. Meanies' miens

35. Suffix with symptom

38. Leadership org. opposed to the G.O.P.

40. Hydrocarbon suffixes

41. Watering hole

44. Speckled horses

47. Start of a Will Rogers quip

51. Worrying sound to a balloonist

52. Fill in the blank with this word: "___ now"

53. They liaise with the FBI

54. Years, to Yves

56. Churchgoer's pet name for his seat?

59. Fill in the blank with this word: "___ flu"

62. Sign of the cross

63. Actress Virna

64. French score

65. To study intensely at the last minute for a test

66. Ltr. extras

67. Gluck's "___ ed Euridice"

68. Zaire's Mobutu ___ Seko

69. They're inflatable

DOWN

1. Arlene and Roald

2. Selfishness

3. Start to like

4. Fill in the blank with this word: "___ spell"

5. Surfaces

6. Shown again

7. Spillane's '___ Jury'

8. Fill in the blank with this word: ""___ the Magician" (old radio series)"

9. Relating to a mystic Jewish sect

10. Suffix with poet

11. Ensured: Abbr.

12. V-J Day pres.

13. Fill in the blank with this word: "___ TomÈ"

21. Wine taster's adjective

22. Fill in the blank with this word: "___ Trueheart of "Dick Tracy""

26. Cut

27. Fill in the blank with this word: "Emerald ___"

28. Sound of a leak

30. SNL' network

31. What a plucker may pluck

32. Fill in the blank with this word: "___ in sight"

35. The basics

36. Spanish aunts

37. Molokai and Maui: Abbr.

39. -

42. Dweller along Lake Volta

43. Fill in the blank with this word: "___ even keel"

45. Vintage

46. They take the bait

48. Hardly optimists

49. with the mouth wide open as in wonder or awe

50. "Romanian Rhapsodies" composer

55. Watch word?

56. Living ___ (what an employer is asked to pay)

57. Words often before a colon

58. 'Waiting for the Robert ___'

59. Cosmetics giant

60. Man, in old Rome

61. Like Brahms's Symphony No. 3

PUZZLE 70

ACROSS

1. Look at

5. View from a beach house

9. Star Wars' creator

14. Singer Green with multiple Grammys

15. Forty-___

16. Wordsworth's Muse

17. Is too good to be true

19. Neighbor of Helsinki

20. Point of success

21. Dulles terminal designer

23. Wagner soprano

25. Went off

26. Nut taken directly from the freezer?

33. Fill in the blank with this word: "Ad ___ (relevant)"

34. Short-billed rail

35. Start of a refrain

36. Some nest eggs

38. Take on

41. Spanish eyes

42. Oohed and ___

44. Philippine island

46. Fill in the blank with this word: "___ alai"

47. Stupendous mentalist!

51. Wall Street deal, in brief

52. Unspecified degrees

53. Fast-paced, slangily

58. The English translation for the french word: tÈtrade

62. Fill in the blank with this word: "___ March, Saul Bellow protagonist"

63. In need of bleeping

65. Soup brand

66. Revenuer

67. Jazz phrase

68. Reserved bar?

69. Puppeteer Tony

70. Sound of a leak

DOWN

1. German "genuine"

2. Fill in the blank with this word: ""___, you!""

3. You and who ___?'

(fighting words)

4. Waterside accommodations provider

5. Informal photograph

6. Web search result

7. Hi-fi spinners: Abbr.

8. Werner ___, 1970's seminar leader

9. Surveys, usually with negative responses

10. Bears: Lat.

11. Fill in the blank with this word: "___ Crunch"

12. Stub ___

13. Fill in the blank with this word: ""So ___?""

18. Winter woes

22. Vodka or whiskey: Abbr.

24. ___ The Magazine (bimonthly with 35+ million readers)

26. Indonesian island

27. Mutual of ___

28. Genre explored by Run-D.M.C. and Aerosmith

29. With 17-Down, a temporary urban home

30. Pilgrim's title

31. Online financial services company

32. What ___ thinking?'

33. Pirate-fighting org

37. Fill in the blank with this word: ""Oh yeah? ___ who?""

39. Hollywood's Bruce or Laura

40. Driving the getaway car for

43. Scott Adams's put-upon comics hero

45. Wedding helper

48. TV's "The ___ Today"

49. Rant and rave

50. Orange dwarfs

53. The Open Window' author

54. Sally ___ (teacake)

55. Worked up

56. Surrealist Joan

57. Fed. lending agency

59. Fill in the blank with this word: ""___ Lama Ding Dong" (1961 hit)"

60. Swedish actor Kjellin et al.

61. Webster's entries: Abbr.

64. Swiss stream

PUZZLE 71

ACROSS

1. Visual way to communicate: Abbr.

4. Fill in the blank with this word: "Dennis Miller book "___, Therefore I Am""

9. Virtual meeting of a sort

14. Fill in the blank with this word: "___-wolf"

15. Fill in the blank with this word: "Brown-___"

16. Mighty Lak' a Rose' composer

17. Fill in the blank with this word: "Big ___ Conference"

18. Danish astronomer Brahe

19. Annual prize won multiple times by Beyonc

20. The English translation for the french word: vieilli

23. "Candid Camera" co-host Jo Ann

24. Like some victories

28. Like non-oyster months

32. The English translation for the french word: ÈgoÔste

33. Ball catcher

36. Really awful, in rap slang

38. Dismissive exclamation

39. Warplane's cargo

43. Fill in the blank with this word: "___ Beach (D-Day site)"

44. Wee

45. Fill in the blank with this word: "i___ boom bah!î"

46. Usual word for an interval when a court suspends business, but doesn't adjourn

49. They may be final or physical

51. Sprightly song

53. Fill in the blank with this word: "___ badge, boy scout's award"

57. Actor Liam's younger kin?

61. The Bee Gees' "How Can You ___ Broken Heart"

64. Worried

65. Verdi's "___ tu"

66. Woolf's "___ of One's Own"

67. Noted Twain portrayer [black]

68. Fill in the blank with this word: "___ el Amarna, Egypt"

69. Lowly ones

70. The ___ Brothers of R & B

71. Fill in the blank with this word: ""___ I Can Make It on My Own" (Tammy Wynette #1 hit)"

DOWN

1. Put ___ to (end)

2. Fill in the blank with this word: "Continental ___"

3. U.S. Open champ, 1985-87

4. Incised printing method

5. Wordsworth's 'Rob ___ Grave'

6. Yiddish writer Sholem

7. ___ Bottling Company (Cleveland fixture for over 85 years)

8. Unscramble this word: torpo

9. Sportscaster with the catchphrase "Oh, my!"

10. Condensation

11. The second Mrs. Sinatra

12. French shooting match

13. West ___ Avenue

21. The English translation for the french word: livrÈe

22. Fill in the blank with this word: "Bill ___, TV's Science Guy"

25. They played Ricky Nelson and Bobby Darin

26. Pancreatic hormone

27. Rail supports

29. She, in S

30. Use a knife

31. Yes or no follower

33. Spanish actress Carmen ___

34. Strength of a solution

35. One of the singing Braxton sisters

37. White Sands Natl. Monument state

40. Quote, part 2

41. Put ___ good word for

42. Crossword grid feature

47. Ticks off

48. Sault ___ Marie

50. Kind of I.R.A.

52. Japanese beer brand

54. Scarlett's love

55. Middle of a famous palindrome

56. Weave, in a way

58. Western Indians

59. Fill in the blank with this word: "Dickens heroine ___ Trent"

60. Local theater, slangily

61. No ___!' ('I give!')

62. We'll teach you to drink deep ___ you depart': Hamlet

63. Fill in the blank with this word: "___'wester"

PUZZLE 72

ACROSS

1. Smallest of the mergansers

5. Fill in the blank with this word: "Easy ___"

10. Fill in the blank with this word: "___ rock (radio format)"

14. Work (out)

15. That is ___...'

16. Uproars

17. Palio di ___ (Italian horse race)

18. The English translation for the french word: rameur

19. Steals, old-style

20. Family financial figure

23. "King Kong" star

24. Veteran journalist

___ Abel

25. The English translation for the french word: datcha

28. Word that means ikindî

30. Facing the pitcher: 2 wds.

34. Pen

36. Wine: Prefix

38. Fill in the blank with this word: "Debussy's "Air de ___""

39. Mathematician's response to 17-, 22-, 51- and 58-Across?

43. T-shirt sizes, in short

44. This, to Th

45. Sibling, often

46. Solar sails material

49. Food preservative: Abbr.

51. Friendly goblin in Scandinavian folklore

52. Hurricane-tracking agcy.

54. U.S.A.F. NCO

56. Gets up for the debate?

62. Fill in the blank with this word: ""Forever, ___" (1996 humor book)"

63. Unscramble this word: tploi

64. Sounds made by 36-Across

66. Zest

67. Neighbor of Pakistan

68. Massachusetts' ___ College

69. Rocky peaks

70. Get rid of

71. West Point rival, for short

DOWN

1. Fill in the blank with this word: "Doo-woppers ___ Na Na"

2. Like some nouns: Abbr.

3. Fill in the blank with this word: "___ perpetua (Idaho's motto)"

4. Odd trinket

5. Venae cavae outlets

6. Woody's wife

7. R.E.M.'s "It's the End of the World ___ Know It"

8. Traveler's guidebook

9. Stage and screen actor ___ Ritchard

10. Used a knight stick on?

11. Start to -matic

12. With 6-Down, 1994 Olympic gold medalist in downhill skiing

13. Sum, ___, fui

21. Sonia of 'Kiss of the Spider Woman'

22. Suffix with neat

25. Voltaire's religious view

26. Song words before gal or shadow

27. Flashlight battery

29. Unwilling

31. Hints

32. The end

33. Truck stop sign

35. S.A.S.E., e.g.

37. Toshiba competitor

40. Pre-election activity

41. The English translation for the french word: conga

42. Words of recognition

47. Year-by-year accounts

48. Towel holder

50. For three: Fr.

53. Fill in the blank with this word: "Easy ___"

55. Went after

56. Library Card Sign-Up Mo.

57. Smartphone introduced in 2002

58. Western Hemisphere abbr.

59. Ye ___ Shoppe

60. The Van Eyck bros. weren't the 1st to use these paints, though sometimes credited with it; they date back much earlier

61. It's in the back row, right of center

65. Winter weather, in Edinburgh

PUZZLE 73

ACROSS

1. Words on a quarter

6. Fill in the blank with this word: "___ 10"

10. Vamp Theda

14. Raider Carl

15. Cause of some impulsive behavior, for short

16. Fill in the blank with this word: "Bric-a-___"

17. Place in the news, 3/28/79

20. Wayne LaPierre's org

21. I could ___ horse!'

22. Zika virus tracker, for short

23. 31-Across activator

27. Fill in the blank with this word: "___ moment"

28. You're in balance if you know it's the complement of "yang"

29. One succumbing to 6-Down

30. Arabian plateau region

32. This is ___'

34. Like smokestacks

38. Start of a playground chant

42. Like some roofs

43. Fill in the blank with this word: ""Are ___ pair?" ("Send in the Clowns" lyric)"

44. Word on a wall, in the Bible

45. Some legal scholars, for short

48. Half a ring

50. Fill in the blank with this word: "___'wester"

51. Chicken dish for Adm. Peary?

55. Tin ___

56. Naldi of old films

57. Tackle-to-mast rope on a ship

58. Loan to a company before it goes public, say

64. Vintage vehicles

65. Mozart's "Dove ___"

66. Chaplin and others

67. Yes-___ question

68. Fill in the blank with this word: "Family ___"

69. Wedding count

DOWN

1. Skedaddle

2. Scot's exclamation

3. Washington's ___ Constitution Hall

4. Fill in the blank with this word: ""___ Fall in Love" (1961 hit by the Lettermen)"

5. Zip

6. Fill in the blank with this word: "Actress ___ Ling of "The Crow""

7. Freud contemporary

8. Bundles

9. Make part of the manuscript

10. Unit of oil production: Abbr.

11. Fill in the blank with this word: ""It's ___ against time""

12. Weary worker's wish

13. Fill in the blank with this word: ""___ Live," 1992 multiplatinum album"

18. Voodoo and wizardry

19. Least wild

23. Trig angle symbol

24. Son of Indira

25. Sicilian resort city

26. Worker in a stable environment?

27. Handle: Fr.

31. Big battery type

33. Wind dir.

35. Terse bridge bid

36. Wooden projection

37. Fill in the blank with this word: "Battle of the ___, opened on 10/16/1914"

39. Singer who spells her name in all lowercase letters

40. Frankie Avalon's '___ Dinah'

41. Safe places

46. Least sweet

47. Take, as an exam

49. I Will Survive' singer Gloria

51. Kitchen implement

52. You can hunt down 2 of the 10 brightest stars in the night sky in this constellation

53. Queeg's command

54. Legendary cowboy ___ Bill

55. Zaragoza's river

59. Brit. honor

60. San Francisco's ___ Valley

61. Storekeeper's stock: Abbr.

62. Power ___

63. Snare

PUZZLE 74

ACROSS

1. The year 1045

5. Windows picture

9. Story, in France

14. Jazz singer Anderson

15. Toodles!'

16. The English translation for the french word: vibraphone

17. Unscramble this word: sgin

18. Saturn vehicles?

19. The Louvre's Salles des ___

20. 1960 Jerry Lewis fairy tale spoof

23. Year in Trajan's reign

24. Player of Det.

Eames on "Law & Order: Criminal Intent"

25. Wrap for a queen

27. The English translation for the french word: yÈmÈnite

30. Maker of pricey bags

32. Financial inst. that bought PaineWebber in 2000

33. Import/export business concern

37. Communication ender

41. Alfresco

42. Wiretapping grp.

43. Subjects of pneumography

44. Virgilian hero

47. The English translation for the

french word: couronne solaire

50. Native village, in South Africa

51. WSW's reverse

52. Part 3 of the quip

58. Sidestep

60. Tiny battery

61. Vertical departure, acronymically

62. Wound up

63. Whole alternative

64. D.O.E. part: Abbr.

65. Per annum

66. Fill in the blank with this word: ""Bill & ___ Excellent Adventure""

67. Tower site

DOWN

1. I.R.S. form 1099-___

2. The Louis whose mother was Marie Antoinette

3. Wood: Prefix

4. Buyer

5. Spanish pianist Jos

6. Fill in the blank with this word: "___ latte"

7. Western Indian

8. The N.Y. Cosmos were in it

9. Principal player in "Grease"

10. Fill in the blank with this word: "Black-throated ___"

11. You can count on them

12. Welcome words to a hitchhiker

13. "Ah, Wilderness!" mother

21. Fill in the blank with this word: ""Ol' Rockin' ___" (bin-mate of the 1957 album "Ford Favorites")"

22. Fill in the blank with this word: "___ hand"

26. Temperate

27. Doctor Zhivago

28. The English translation for the french word: ÈbÈne

29. U.S.M.C. E-8

30. Unleashes

31. Yield, as interest

33. Suitable for teen

audiences

34. Nary ___

35. Radio station in a 1970 Paul Newman title

36. Union and others: Abbr.

38. Craig of the N.B.A.

39. Investigator who finds someone's birth mother, say

40. Chaplin and others

44. Starting lineups

45. Suffix with Mozart

46. Football squad

47. Jai alai basket

48. Sounding right, in a way

49. Marie Antoinette, e.g.

50. Unflashy

53. Where hops are dried

54. Sham

55. Fill in the blank with this word: "Emporio ___"

56. Ways: Abbr.

57. Nickelodeon's "___ the Explorer"

59. Mandela's land: Abbr.

PUZZLE 75

ACROSS

1. This fermented honey-&-water beverage was a favorite of Chaucer's miller & of the god Thor

5. Skin Bracer alternative

9. Start of an opinion

14. Lenin's "What ___ Be Done?"

15. TV screen meas.

16. "Butter knife" of golf

17. Vessel with a load

18. Vissi d'___' (Puccini aria)

19. Donated

20. Weekend ice cream treat?

23. Japanese vegetable

24. Wisc. clock setting

25. Movie buff: Var.

29. What a cedilla indicates

31. John ___-Davies of the "Lord of the Rings" trilogy

33. Geometric figs.

34. It was once divided

36. Like British bishops

39. Gave Grey Poupon to the head of the table?

42. Player with a string quintet

43. Special people

44. Verdi's "___ tu"

45. Vermont but not New Hampshire, e.g.?

47. Fill in the blank with this word: ""___ it!" ("Didn't fool me!")"

51. Setting for Longfellow's "The Wreck of the Hesperus"

54. Mandela's land: Abbr.

56. I ___ Rock'

57. Dieter's credo?

60. Plant with two seed leaves

63. 1987 Suzanne Vega hit

64. Finish this popular saying: "Hard work never did anyone any____."

65. Fill in the blank with this word: "'"Y'know what ___?""

66. Words with a nod

67. Use a knife

68. Sidewalk ___

69. Peloponnesian War participant

70. Fill in the blank with this word: "___ Fjord"

DOWN

1. Wife, informally

2. Pre-euro Portuguese currency

3. Untold

4. Stewart's role in "Harvey"

5. Fill in the blank with this word: "All in ___ work"

6. The industrialized nations

7. about three feet long exclusive of tail

8. Spook's employer, with "the"

9. Tom Harkin, for one

10. Pfeiffer of TV's 'Cybill'

11. S, to a frat guy

12. Ticket abbr.

13. Robert Burns's "___ Wild Mossy Mountains"

21. With 16-Across, way leading to a highway

22. Something to fall into

26. Study of the stars: Abbr.

27. Wallpaper meas.

28. Terse reproof

30. Shortening used in recipes

32. Words mouthed to a TV camera

35. The English translation for the french word: rÈtine

37. Of a heart part

38. What to do

39. Professional prefix

40. Take ___ (swing hard)

41. Complete

42. TV's "___ Ramsey"

46. Land south of Hadrian's Wall

48. Some speech sounds

49. Fill in the blank with this word: ""The Essence of ___," Food Network show"

50. Start to like

52. W.W. II vessel

53. UnitedHealth rival

55. Dairy Queen offerings

58. To remove, as from office

59. Fill in the blank with this word: ""___ that's your game!""

60. Trash

61. The Monkees' '___ Believer'

62. The 21st, e.g.: Abbr.

PUZZLE 76

ACROSS

1. Fill in the blank with this word: ""___ to recall "

6. Twice tetra-

10. Fill in the blank with this word: "___ McAn shoes"

14. You've got the wrong guy!'

15. Essential company figure

16. Philosopher David

17. Like aprons, at times

19. They're often bitter

20. Words from a would-be protégé

21. Turn over again

23. Lummoxes

25. The second part

missing in the author's name ___ Vargas ___

26. Famous place with a hint to this puzzle's theme

32. Skeleton components

33. Wine taster's adjective

34. Meditation sounds

37. Heroin, slangily

38. "Hamlet" courtier

40. "That was close!"

41. Elev.

42. Ural River city

43. Fill in the blank with this word: "2003 Nick Lachey hit "___ Swear""

44. 2001 50-Across nominee

47. Mirabile ___ (wonderful to say)

50. Shakespearean king

51. Library section

54. Fire sources

59. I Lost It at the Movies' writer Pauline

60. Some clerks

62. One way to race

63. Sound system brand

64. Tennessee ___ Ford

65. What knows the drill, for short?

66. Dilbert co-worker

67. Marie Antoinette, e.g.

DOWN

1. Research facility: Abbr.

2. 1982 title role for Meryl Streep

3. Ici ___ (here and there, to Th

4. Old Apple product marketed to schools

5. Ways

6. Political activist James known for undercover videos

7. Nav. leader

8. Six-foot vis-

9. French wave

10. Fill in the blank with this word: "Buddy Holly's "___ Be the Day""

11. 1950s fad item

12. Biblical dry measures

13. Legendary Washington hostess Perle ___

18. Yao Ming teammate, to fans

22. Wow

24. View coral reefs, maybe

26. Talk effusively

27. Cockney greeting

28. Sci. course

29. This 3-letter word means to pester or badger; it can also refer to a worthless horse

30. Mai ___

31. QB Manning

34. Fill in the blank with this word: "___ Valley Conference"

35. The Bible Tells ___'

36. Triathlon leg

38. A Thing ___' (Beach Boys song)

39. Secure online protocol

40. Sentence part: Abbr.

42. Fill in the blank with this word: "___ probandi"

43. First-rate

44. Slanted type

45. Trim

46. Rabbit ___

47. What is the capital of this country - Senegal

48. Give ___!' ('Try!')

49. Traction provider

52. Fill in the blank with this word: "___ noche (tonight, in Tijuana)"

53. Writer's supply: Abbr.

55. Clinton's #2

56. Present opener?

57. Teutonic turndown

58. Snick-or-___

61. Philosopher ___-tzu

PUZZLE 77

ACROSS

1. V preceder

5. Town line sign abbr.

9. Old draft deferment category for critical civilian work

13. How football's Jerry was addressed as a boy?

16. Lux. neighbor

17. Manages to get through

18. Writer's Market abbr.

19. Time-honored Irish cleric, for short

20. Wiped out

22. Wormer, say

23. Lived ___ (celebrated)

25. Square, in 1950s slang, indicated visually by a two-hand gesture

27. Travelers' papers

30. Stamps, say

32. Sue Grafton's "___ for Alibi"

33. Scattered, as seed

34. Wayne LaPierre's org

35. It may be blacked out

38. Maritime CIA

39. Star born Frederick Austerlitz

41. Lb. or oz.

42. The English translation for the french word: posada

44. Fill in the blank with this word: "Feather ___"

45. Title page?

46. Fill in the blank with this word: ""The ___" (Uris novel)"

47. With 100-Across, Naples opera house Teatro di ___

48. Etc. and ibid., e.g.

49. Declare, old-style

51. He played Mowgli in "Jungle Book"

53. Indonesia's ___ Islands

54. Asian goat

56. Fill in the blank with this word: ""I Am ... ___ Fierce," #1 Beyonc"

59. The Godfather' co-star

61. Apparatus named for a French physician

64. This instrument of the cult of Apollo lent its name to the type of poetry it accompanied

65. Exhibit artfulness

66. "As we have therefore opportunity, let ___ good to all men": Galatians

67. Undesirable serving

68. Way from Syracuse, N.Y., to Harrisburg, Pa.

DOWN

1. White House inits.

2. Writer's supply: Abbr.

3. U.S.A.F. NCO

4. Perfect but impractical

5. Verdi baritone aria

6. Veracruz Mrs.

7. Small songbirds

8. Some batteries

9. USA alternative

10. Tries to trap something

11. TV's ___ twins

12. Time for potty training, maybe

14. The Louvre's Salles des ___

15. Go-aheads

21. Zigzag

24. Parallel ___ Moresby

26. Bookie's charge, for short

27. Letters on a R

28. Prefix with sphere

29. Protection against rustling?

31. Most sacred

34. Wiretapping grp.

35. Fill in the blank with this word: "___-la-la"

36. Hate or fear follower

37. Waiting area announcements, briefly

39. Like some professors

40. Water's conductivity comes from these particles, like positive sodium ones & negative chlorine ones

43. Fill in the blank with this word: ""___ approved" (motel sign)"

45. Black & Decker offering

47. Straight run for skiers

48. Lack of restraint

49. Strongly hopes

50. Unilever brand

52. There's many ___ 'twixt the cup and the lip'

53. Since 1920 the organization known by these 4 letters has helped defend the rights & freedoms of our people

55. Cambodian money

57. School subj.

58. Handle: Fr.

60. Fill in the blank with this word: "___-noir (modern film genre)"

62. Band with the 1999 hit "Summer Girls"

63. Width measure

building in Islam

PUZZLE 78

ACROSS

1. Fill in the blank with this word: "2000 Olympic hurdles gold medalist ___ Shishigina"

5. West Virginia resource

9. Gave up

14. Fill in the blank with this word: "___ oak"

15. Fill in the blank with this word: "___ in a blue moon"

16. Fill in the blank with this word: "___ fixes"

17. Winter coating

18. Michelle of "Crouching Tiger, Hidden Dragon"

19. Actress Beulah

20. Hoosier pro

23. Formal hat, informally

24. Soldier

25. Workout unit

28. They test reasoning skills: Abbr.

30. Ordered group of numbers in math

32. Permitted

33. Start of instructions for solving this puzzle

37. Mussorgsky's 'Pictures ___ Exhibition'

39. Power ___

40. Heads ___, tails...'

41. Fill in the blank with this word: ""Merci beaucoup" : France :: ___ : Japan"

46. Fill in the blank with this word:

"Conductor ___-Pekka Salonen"

47. Houston pro soccer team

48. Welcome

50. Wilt

51. Alley org.

53. The English translation for the french word: ouvrage

54. 50's-90's jazz singer

59. Successor to Clement VIII

62. Unstable leptons

63. Sound

64. Salami choice

65. Penny-___ (trivial)

66. Helgenberger of "CSI"

67. Proust title character

68. Old Chinese money

69. Some TV spots, briefly

DOWN

1. It's you! What a surprise!'

2. Fill in the blank with this word: "Crazy as a ___"

3. Unscramble this word: ladg

4. Hindu drink of the gods

5. the affectation of being demure in a provocative way

6. Two semesters

7. Reynolds film "Rent-___"

8. Composer Franz

9. Symbol on an old quarter

10. Whiff

11. Fill in the blank with this word: "China's Sun Yat-___"

12. Turner of TV channels

13. Fu-___ (legendary Chinese sage)

21. Rental units: Abbr.

22. Water-to-wine site

25. Made like a geyser

26. usually good-naturedly mischievous

27. Fill in the blank with this word: ""Lovergirl" singer ___ Marie"

28. One of the Jacksons

29. With force and much noise

31. Switzerland's Bay of ___

32. Hollywood's Alan and Diane

34. Maritime CIA

35. Scott Joplin's "Maple Leaf ___"

36. Worker with books, for short

38. Fill in the blank with this word: ""...___ thousand times..."″

42. Semiterrestrial organism

43. The ___ Report (upscale magazine)

44. 38-Across's real name, in brief

45. Mechanic's ___

49. How some kids spend the summer

52. Old tombstone abbr. meaning "at the age of"

53. Instrument with fingerholes

54. White matter component

55. Indian of the Sacramento River valley

56. Harem rooms

57. Peseta : Spain :: ___ : Italy

58. Yuletide quaffs

59. Some TVs and smartphones

60. Repugnant exclamation

61. Three ___ match

PUZZLE 79

ACROSS

1. I...
4. All excited
8. Boat with an open hold
12. Bunk
15. "Rocks"
16. Hindu Mr.
17. Voltage through motion
19. Chinese restaurant offering
21. Logician
22. Begin
23. Extreme oldness
25. "___ Gang"
26. Coal carrier
27. Pandowdy, e.g.
28. Absorbed, as a cost
29. Productive period
32. 2004 nominee
34. Celebrations
36. Informal term
39. Incorrect in behavior
40. Branch
41. Climbing bean or pea plant
42. Antiquity, in antiquity
43. Certain protest
45. Blubber
46. "ER" network
48. Boat in "Jaws"
52. 1999 Pulitzer Prize-winning play
54. See-through sheet
57. Long, long time
58. Bud holders?
61. Aggressive
63. Animal hides
64. "Go on ..."
65. Field
66. "___ we having fun yet?"
67. Amigo
69. ___ Today
72. Modern F/X field
73. straight line or lines
76. 1,000 kilograms
79. Power and authority
80. Commits sabotage
82. Security checkpoint
83. Antares, for one
84. Schuss, e.g.
85. Oolong, for one
86. Change
87. Cravings
88. "Comprende?"

DOWN

1. Butts
2. "God's Little ___"
3. ...
4. Eccentric
5. Gangster's gun
6. Witchcraft and sorcery
7. Sea birds; used as fertilizer
8. ...
9. Annoying and unpleasant
10. Certain Arab
11. Inefficient in use of time and effort and materials
12. Narrowly triangular, wider at the apex
13. Clytemnestra's slayer
14. Minimally worded
18. Elephant's weight, maybe
20. Ladies' bag
24. Kind of drive
29. Org. that uses the slogan 'Aim High'
30. Legal prefix
31. Gets rid of
33. Administrative unit of government
35. Edible starchy
37. "I" problem
38. Irish Republic
39. The terminal section of the alimentary canal
44. Blockhead
47. High school class, for short
49. Inclination
50. Chanel of fashion
51. Again
53. Who stimulates and excites people
55. Excessive
56. A muscle which raises any part
58. Most economical
59. Phormio' playwright
60. Euripides drama
62. A homosexual man
64. Parallel or straight
68. "Home ___"
70. Flip, in a way
71. Winged
74. Not just "a"
75. "I, Claudius" role
77. Microwave, slangily
78. Ashtabula's lake
81. Blackout

PUZZLE 80

ACROSS

1. People person

6. Bro

10. Chill

14. Charm

15. ___ probandi

16. Dismal

17. Disappointment after good expectations

19. Fraction of a newton

20. Bro, for one

21. Boot

22. "Much ___ About Nothing"

24. A person who belongs to the political left

26. Capsule contents in a spy movie

30. Coarse file

31. Dangerous dive

32. A homosexual person

34. Frost lines

35. Ground cover

36. Furnace output

37. Cache

38. ...

39. "Catch-22" pilot

40. Architectural projection

41. Coaster

42. Boundary

44. ___ Minor

46. Testimony in a court of law

47. Results of blizzards

49. All the rage

50. Endorse

51. "Concentration" pronoun

53. Brio

56. Country established in 1948

59. Doctor Who villainess, with "the"

60. Ashtabula's lake

61. Not suitable

62. "Beowulf," e.g.

63. Angry, with "off"

64. Cheesy sandwiches

DOWN

1. Bounders

2. In-box contents

3. A workplace where lumber is

4. Clairvoyance, e.g.

5. Intoxicates

6. Drench

7. Condo, e.g.

8. ...

9. A writer

10. Extras

11. Chipper

12. Ashes holder

13. Barely get, with "out"

18. Bon mot

23. "Over" follower in the first line of "The Caissons Go Rolling Along"

25. Certain protest

26. Christmas ___

27. On or relating to the same side

28. Electron tube

29. Broke off

31. Crow's home

32. Hamlet's father, e.g.

33. Cliffside dwelling

34. Full of veins; veined

37. Large fishnet

38. "___ I care!"

40. Assortment

43. Italian, e.g.

44. Bow

45. A geographical area

47. No longer in

48. Used a broom

50. Cheese on crackers

52. Beanery sign

53. "... ___ he drove out of sight"

54. Drink from a dish

55. "Gimme ___!" (start of an Iowa State cheer)

57. Bauxite, e.g.

58. "___ moment"

PUZZLE 81

ACROSS

1. Fill in the blank with this word: "___ loaf"

6. Fill in the blank with this word: "Arthur C. Clarke's "Rendezvous With ___""

10. One of the Aleutians

14. You can hunt down 2 of the 10 brightest stars in the night sky in this constellation

15. Whiff

16. Widespread

17. Fill in the blank with this word: ""___ your life!""

18. TV actor Katz

19. Fill in the blank with this word: "60's TV's ___ May Clampett"

20. Like a line

23. Sue Grafton's '___ for Ricochet'

24. It was split in 1948: Abbr.

25. Taiping Rebellion general

26. Terre Haute sch.

29. Woman's loose-fitting garment with flared legs

32. Provincial capital in the Dominican Republic

35. Olympic sprinter ___ Boldon

36. Tidal flood

37. Waffle House alternative

38. Hamilton' actress ___ Elise Goldsberry

41. Word before window or end

42. Yemen's capital

44. Tin ___

45. Assns. and orgs.

46. Traffic sign that indicates a possible temporary road closure

50. Whitman's "A Backward Glance ___ Travel'd Roads"

51. Long-running film role

52. End of many an E-mail address

53. Where "48 Hours" airs

56. Tie up a Midwest senator?

60. Fill in the blank with this word: "Beethoven's "Archduke ___""

62. Shall I compare ___ to a summer's day?'

63. High-altitude home

64. Quaint affirmative

65. Fill in the blank with this word: "___-poly"

66. Fill in the blank with this word: ""Bonne ___!" (French cry on January 1)"

67. Fill in the blank with this word: "___-tiller"

68. Play to ___ (deadlock)

69. Stock options?

DOWN

1. Unscramble this word: ohrno

2. Rice-___

3. The Chi-___ (1970s R & B group)

4. Fill in the blank with this word: "___ pyramid, four examples of which are seen in this puzzle"

5. Golf's Sorenstam

6. Suite door posting

7. Wearers of four stars: Abbr.

8. St. ___, Switzerland

9. Vocal style

10. Gray ___

11. Notorious 1920's criminal

12. Workers' grp. founded 1886

13. Fill in the blank with this word: "Church ___"

21. The English translation for the french word: molaire

22. Yep's opposite

27. Watch part

28. Fill in the blank with this word: "___ manual"

29. Tree with oblong leaves

30. Spaced (out)

31. Exasperation exclamation

32. Bungle

33. Second City's #1 airport

34. Swindler

39. Scout's skill

40. Rhone feeder

43. Leigh Hunt's "___ Ben Adhem"

47. Talking points?

48. Doesn't work anymore, informally

49. Virgilian hero

53. Bony part

54. Remove the dirt from?

55. Dairy Queen offerings

57. QB Tony

58. Like Christmas in Madrid?

59. Fill in the blank with this word: "1992 Heisman winner ___ Torretta"

60. Norse war god

61. Rock's ___ Speedwagon

PUZZLE 82

ACROSS

1. Take ___ at (try)

6. Omar of "The Mod Squad," 1999

10. UV blockage nos.

14. State in Brazil

15. Soft roe

16. Classic Nestl

17. Fill in the blank with this word: "___ Line (German/Polish border)"

19. Sell short

20. The English translation for the french word: RAU

21. What's tapped at a beer bust

22. Violent, perhaps

24. Some brass

28. Fill in the blank with this word: ""___ beam up" ("Star Trek" order)"

30. Purpose

31. Knuckler alternative

32. This one's ___'

33. Prefix with fauna

36. Whiskies

37. Tragic James Fenimore Cooper character

39. Place for keys and lipstick

40. Unrealized 60's Boeing project

41. Summer Games org.

42. Brewer Adolphus

43. Smashes from Sampras

45. Where ___

46. *One who's often doing favors

50. Writers Shreve and Brookner

51. Military asst.

52. Barges

55. Pro ___ (proportionately)

56. What it's like to be Spider-Man?

60. Many a holiday visitor / Bandit

61. Webzine

62. Trig angle symbol

63. Fill in the blank with this word: "___ Vista"

64. Trails off

65. Fill in the blank with this word: ""Super Duper ___" (anime series)"

DOWN

1. Leigh Hunt's "___ Ben Adhem"

2. Thompson of "Family"

3. having the property of becoming permanently hard and rigid when heated or cured

4. You can bob for apples because they're 25% this, which allows them to float

5. Withdrawal carrying a steep penalty?

6. Solzhenitsyn, e.g.

7. TV's Magnum and others

8. Some nouns: Abbr.

9. Breastbones

10. Plaza

11. Fill in the blank with this word: "___ Arenas, port in 93-Down"

12. Cuba's Castro

13. Variety listings

18. Want ad inits.

23. Adjust, as a clock

25. Lawyers: Abbr.

26. Trattoria dumplings

27. Writer ___ St. Vincent Millay

28. Bank-to-bank transactions: Abbr.

29. House Committee on ___ and Means

33. Auto-stopping innovation

34. Fill in the blank with this word: "Explorer Cabeza de ___"

35. See 103-Across

37. Assessment paid only by those who benefit

38. Norse goddess of fate

39. Tutsi foe

41. Part of the eye

42. Pumas, e.g.

43. What is the capital of this country - Canada

44. Fill in the blank with this word: ""Lowdown" singer Boz ___"

46. Santa ___

47. With everything counted

48. Join, as a table

49. When some summer reruns are broadcast: Abbr.

53. Fill in the blank with this word: "Baseball's ___ Gaston"

54. Flies away

57. Pres. appointee

58. Snatch

59. Moo ___ pork

PUZZLE 83

ACROSS

1. Where Moses got the Ten Commandments

8. USA alternative

11. What's funded by FICA, for short

14. Fill in the blank with this word: "English poet Coventry ___, who wrote "The Angel in the House""

15. Fill in the blank with this word: "___ in apple"

16. Stamp not reqd.

17. Surf serving #4

19. Source of some rings

20. Per ___ (daily)

21. World production of this is now 84 million barrels a day, up about 10 million barrels from 1997

22. Wood: Prefix

23. 2000 site

27. Way from Syracuse, N.Y., to Harrisburg, Pa.

28. 1965 #1 hit by the Byrds

29. Fill in the blank with this word: ""If ___ Would Leave You""

30. Word on a dipstick

31. Fill in the blank with this word: "Architect ___ Ming Pei"

33. Some hip-hop women

34. the hole in a woodwind that is closed and opened with the thumb

36. Phylicia of stage and screen

39. Chopin's "Butterfly" or "Winter Wind"

40. The English translation for the french word: viscÈral

43. Work translated by Pope

44. Windsor's prov.

45. Natal native

46. Unwelcome sight in the mail

50. Henley Regatta setting

51. Groove-billed ___

52. Fill in the blanks with these two words: ""Whatcha ___?""

53. Sir ___ McKellen (Gandalf portrayer)

54. Spanish sherry

58. Yabba dabba ___!'

59. What's missing from a KO?

60. Patella

61. Most miserable hour that ___ time saw': Lady Capulet

62. Reply facilitator: Abbr.

63. Snags

DOWN

1. UK legislators

2. Tit for ___

3. Star-___

4. Nothing's broken!'

5. Phrase in some apartment ads

6. Fill in the blank with this word: ""You ___ Lucky Star" (1935 #1 hit)"

7. Weapon for Iraqi insurgents: Abbr.

8. One who knows "the way"

9. Call that may complete a full count

10. Wind dir.

11. You may want to stop reading when you see this

12. Member of a 1990s pop quintet

13. Shelley poem

18. Home of Notre Dame

22. Fill in the blank with this word: ""C'est ___""

23. Veracruz Mrs.

24. Wall Street earnings abbr.

25. Mr. ___, radioactive enemy of Captain Marvel

26. Fill in the blank with this word: "En ___ (by the rules: Fr.)"

31. Fill in the blank with this word: ""If ___ nickel...""

32. Fix a squeak

33. Fill in the blank with this word: ""What a ___!" (beach comment)"

34. Rodgers and Hart's "___ Love"

35. Tutsi foe

36. Swimmer's fear

37. Song written by Queen Liliuokalani

38. Yes sir!,' south of the border

40. Wild llama

41. Where Einstein was born

42. Ram, in Ramsgate

44. Wife of Paris, in myth

45. They're great on Triple Letter Scores

47. This type of radiation, & letter of the alphabet, hulked out Bruce Banner

48. Patsy's pal on "Absolutely Fabulous"

49. Undersides

54. The Unsers of Indy

55. Damascus's land: Abbr.

56. Peace Nobelist Kim ___ Jung

57. Special ___

PUZZLE 84

ACROSS

1. With ___ of thousands'

6. Limerick's rhyme scheme

11. White lie

14. The English translation for the french word: suiveur

15. Set aside

17. Hard to hold

19. Taxonomic suffix

20. Informational symbol

21. Tale of adventure

22. Fill in the blank with this word: "___ sponte (of its own accord, at law)"

23. Sheik ___ Abdel Rahman

26. Fill in the blank with this word: ""Tuesday ___" (Count Basie tune)"

29. (of taxes) increasing as the amount taxed increases

34. Temple

35. Go to ___

36. Writer St. John ___

37. They're debatable

40. Vocal style

41. O.T. book

42. Obi accessory

43. Singer of the lyric formed by the ends of the answers to the four starred clues

45. Illinois city, site of the last Lincoln-Douglas debate

46. Swamp ___

(predatory fish)

47. Loos

48. Fill in the blank with this word: "___ la Douce"

52. Fill in the blank with this word: "___ Pendragon, King Arthur's father"

54. Triple-decker, perhaps

57. 1974 Rolling Stones hit

61. #4

62. Good, to Guido

63. Suffixes with glycer- and phen-

64. "Top Chef" host Lakshmi

65. Magical symbol

DOWN

1. Fill in the blank with this word: ""I know not why I ___ sad": Shak."

2. Worshiper of the sun god Beal

3. Play to ___ (deadlock)

4. Sponge (up)

5. Tom Cruise movie

6. Well-ventilated

7. "I'll take whatever help I can get"

8. The Creator, to Hindus

9. Would-___ (aspirants)

10. White ___ ghost

11. Finish this popular saying: "The best things in life are_____."

12. Folk singer Burl

13. Whack

16. School subj.

18. She, in S

22. the opposite of often

24. Take away, at law

25. Onion relative

26. Fill in the blank with this word: ""There's no such thing ___ publicity""

27. Complete, informally

28. There was gold in the hills of this host city as the Times noted the start of the Winter Olympics there on Feb. 10

29. Fill in the blank with this word: ""Well, ___ darn!""

30. Profits

31. Fill in the blank with this word: ""___ beauty, so to speak, nor good talk ...": Kipling"

32. When to say "Feliz A

33. Adores

35. Working stiff

38. Verb ending

39. Bhikkhuni : Buddhism :: ___ : Catholicism

44. Sidestepped

45. Tumblers, e.g.

47. United

48. Fill in the blank with this word: ""___ Anything" ("Oliver!" song)"

49. Vex

50. Spanish ___

51. Military asst.

53. Fill in the blank with this word: "Boxer Oscar De La ___"

54. Retriever

55. TV's Anderson

56. Unscramble this word: loto

58. Fill in the blank with this word: "Alley ___"

59. The plural of the word ovum

60. Arles assent

PUZZLE 85

ACROSS

1. Yearn

7. Hurdle for some univ. seniors

11. WWII intelligence org

14. Taken for

15. Their, in Munich

16. Unscramble this word: alp

17. Steely Dan hit, 1980

19. Photography abbr.

20. Fill in the blank with this word: ""___ Como Va" (1971 Santana hit)"

21. Superman foe ___ Luthor

22. Having flow controls

24. What nomads do

26. The English translation for the french word: soma

28. Fill in the blank with this word: "Amazon ___"

29. Scott who wrote "Presumed Innocent"

31. Worn out

33. Native village, in South Africa

34. Worse in quality, slangily

38. The English translation for the french word: moi

39. See eye-to-eye?

43. Stationer's item: Abbr.

44. Annual parade subject

45. Without ___ to stand on

47. Year of Bush's swearing-in

48. Marine creature with a transparent, saclike body

52. Like Ibsen, to his countrymen

54. Germany's ___ von Bismarck

57. "Golly!"

58. Language of central Mexico

60. Org. with an annual televised awards ceremony

62. Star of "Youngblood," 1986

63. Spanish queen until 1931

64. Bygone era, which will help answer the five capitalized clues

67. RR stop

68. Web ___

69. Violent, perhaps

70. Turf

71. Novelist Seton

72. Trials

DOWN

1. Middle of the riddle

2. With 27-Across, holiday celebrators' farewell

3. Annually

4. The Admiral Benbow ___ ("Treasure Island" locale)

5. Unscramble this word: liar

6. Old laborers

7. Sch. on the Charles

8. Thank Heaven for Little Girls' singer

9. Gray ___

10. Wt. of some flour sacks

11. Shrub also known as Russian olive

12. The old man in "The Old Man and the Sea"

13. Magnificence

18. Skeleton's head?

23. Fill in the blank with this word: "___ deferens"

25. Ridicule

27. Frenzied place at a rock club

30. Zachary Taylor, for one

32. Fill in the blank with this word: "___-pointe (ballet position)"

35. City in Judah

36. Information often set in brackets

37. Relieves (of)

39. Quality of good ground beef

40. Being borrowed by

41. Like tennis serves

42. Widen, in a way

46. Sch. in Atlanta

49. Tease

50. "Rats!"

51. What the remorseful might make

53. Phi Beta ___

55. Cable TV giant

56. Old-fashioned card game, in Britain

59. Fill in the blank with this word: "Air___, discount carrier"

61. They're what a pompous person "puts on"

65. Isn't ___ bit like you and me?' (Beatles lyric)

66. Fill in the blank with this word: "___ in Thomas"

PUZZLE 86

ACROSS

1. Mineral residue

5. Spelling champ?

10. Nursery school, informally

14. N.B.A.'s Nick Van ___

15. State in Brazil

16. Ancient Semitic fertility goddess

17. Popular basketball shoe

20. Rock's Brian ___

21. The Bible Tells ___'

22. Pancreatic hormone

23. "My stars!"

24. Block

26. raid and rove in search of booty

29. Spear

30. Fill in the blank with this word: ""What a ___!" (beach comment)"

33. Fill in the blank with this word: "___ and Thummin (Judaic objects)"

34. The English translation for the french word: taxer

35. The Monkees' '___ Believer'

36. LIGHTS!

40. Topper

41. Spanish soccer star Sergio ___

42. Wrist-elbow connector

43. Dutch city

44. Old Soviet secret police org.

45. Welcomed, as the New Year

47. Victory: Ger.

48. This sac in a bird's egg is there to provide nourishment to the embryo

49. Give more cushioning

52. White House's ___ Room

53. Fill in the blank with this word: "Agatha Christie's "The ___ Murders""

56. Genre of 17- and 56-Across

60. You can get a charge out of it

61. One Word ___ Often Profaned' (Shelley poem)

62. Word-of-mouth

63. Striking end

64. Valuable find

65. The Pekingese, Chihuahua, & Tonka's "Pound Puppies" fall under this classification of dogs

DOWN

1. Succumb to mind control

2. White matter component

3. One who follows the news

4. Year in Claudius's reign

5. Conventional

6. Deadlock

7. Lt. Kojak

8. World Factbook publisher, in brief

9. Lyricist ___ David

10. Pari ___ (fairly)

11. V preceder

12. Useful Latin abbr.

13. Russell of 'Running Wilde'

18. Flightless bird: Var.

19. Epsom salts

23. Lady of Spain

24. Contemptible sneaks

25. Jai ___

26. Poor orator, perhaps

27. Sure competitor

28. Fill in the blank with this word: "___ Oro"

29. Fill in the blank with this word: ""All I Ever Need ___" (Sonny & Cher hit)"

30. Microsoft chief, to some

31. Yemeni's neighbor

32. Suez Crisis figure

34. Campus buildings

37. Some Tatooine workers

38. Living ___ (what an employer is asked to pay)

39. The Sex Pistols' genre

45. Rod

46. Voice below soprano

47. East Indian heartwood

48. Comic Smirnoff

49. Tell the host yes or no

50. Veteran journalist ___ Abel

51. Matador's move

52. Fill in the blank with this word: "___-1 ("Ghostbusters" vehicle)"

53. Old magazine ___ Digest

54. Present time, for short

55. V-8's eight: Abbr.

57. Fill in the blank with this word: "Black-throated ___"

58. Where Oskar Schindler is buried: Abbr.

59. Fill in the blank with this word: "Bon ___"

PUZZLE 87

ACROSS

1. Yom Kippur service leader

6. Repeated interjection in the Rolling Stones' "Miss You"

10. Union foe

14. Fill in the blank with this word: "___ the hole"

15. Fill in the blank with this word: ""___-daisy!""

16. Roman statesman ___ the Elder

17. PBS policy

18. Sub

19. Fill in the blank with this word: "Astronomy's ___ cloud"

20. Classic Miles Davis album ... or a hint to the start of 17-, 22-, 37- or 45-Across

22. They often precede la's

23. *Group with the 2000 #1 hit "It's Gonna Be Me"

24. Tiny time unit: Abbr.

26. Laid bets at a casino

29. Cello feature

33. Part of USPS: Abbr.

37. Whack

38. Detailed, old-style

39. Fill in the blank with this word: "___ Cologne (skunk of old cartoons)"

40. Abbr. after Ted Kennedy's name

42. Unscramble this word: gang

43. Some October babies

45. Poker supplies

46. With 8-Down, source of an ethical dilemma

47. Wordless song: Abbr.

48. Under-the-sink fitting

50. Singer Brickell

52. Fill in the blank with this word: "___ artery"

56. Vestments, e.g.

59. Republication

63. Tabby talk

64. Fill in the blank with this word: ""The House Without ___" (first Charlie Chan mystery)"

65. Vulcan portrayer

66. Trollope's "Lady ___"

67. Fill in the blank with this word: "___ East"

68. Need for the winner of a Wimbledon men's match

69. Union member

70. White-tailed eagle

71. Successful job applicant

DOWN

1. What stripes may indicate

2. Flip ___ (decide by chance)

3. You don't know ___'

4. Fussbudget

5. One way to break out

6. Bushman's home

7. Volkswagen competitor

8. Royal fern

9. Pumps up

10. Sausage-wrapped British breakfast dish

11. Openness

12. Gillette ___ Plus

13. Nonhuman co-hosts of TV's "Mystery Science Theater 3000"

21. King and queen

25. Red ___ (young amphibian)

27. Open a New Window' musical

28. Difference in days between the lunar and solar year

30. Sheik ___ Abdel Rahman

31. Watercolorist ___ Liu

32. Wired

33. Songs for one

34. Fill in the blank with this word: "___ chief (publ. honcho)"

35. You've got 24 of these in the front of your chest, protecting your inner organs

36. Facet joints connect them

38. See 34-Across

41. White Sulphur ___, W. Va.: Abbr.

44. Suffix with tank

48. Cheater, perhaps

49. Statue base

51. Witless

53. You're looking at him!'

54. Archer, at times

55. Robert of Broadway's "My Fair Lady"

56. Transcript stats

57. West End classic "Charley's ___"

58. Cell stuff that fabricates protein, for short

60. Withdraw gradually

61. Fill in the blank with this word: "Brontí's "Jane ___""

62. Where Loews is "L"

PUZZLE 88

ACROSS

1. Worked on Broadway

6. Some hairstyles

10. Table salt, to a chemist

14. French versifier

15. Qatar's capital

16. Pas ___ (gentle ballet step)

17. Obliged out of integrity

19. Fill in the blank with this word: ""___ Stars," #1 hit for Freddy Martin, 1934"

20. Press agent?

21. Manute ___ of basketball

22. Fill in the blank with this word: ""___ of the Flies""

23. Ukr. and Lat., once

24. Opium maker

26. Techno-funk band with the 1991 #1 hit "Unbelievable"

28. Wore away

29. Three-toed critters

33. www page creation tool

37. Zap in the microwave

39. Narrow, in a way

40. Specious reasoning

42. Resets

44. Nymph in Greek myth

45. Grape for winemaking

47. Writer's supply: Abbr.

48. Torments

50. Taxonomic suffix

51. Wall Street org.

52. Fill in the blank with this word: "___ Tamid (synagogue lamp)"

53. Zaragoza's river

57. Medieval chest

60. To the ___ power

62. Some people use spray deodorants, but many conservationists prefer this non-aerosol kind

64. Fist ___ (modern greeting)

65. Cartoons collected in "Cows of Our Planet"

67. Fill in the blank with this word: ""No ifs, ___ ...""

68. Spy Mata ___

69. Quinn of 'Legends of the Fall'

70. Prefix in hematology

71. Priory of ___, group in "The Da Vinci Code"

72. Type of question

DOWN

1. Sap-sucker's genus

2. The Rockies' ___ Field

3. Wagner's Tannh

4. Wide collars

5. Fill in the blank with this word: ""Hello ___" (old comedy intro)"

6. Japanese vegetable

7. Certain ranch name ... or this puzzle's theme

8. Speed skater Apolo Anton ___

9. Fill in the blank with this word: "___'s Wells (renowned London theater)"

10. Providers of tip-offs?

11. Fill in the blank with this word: "___-ran"

12. Woodwind instrument: Abbr.

13. Smutty

18. William Jennings ___ "Cross of Gold" speech

25. The thing is what a lumberjack leaves behind; the place is where a politician speaks

27. Electrical units

30. Zola's streetwalker

31. Beginning

32. Series of legis. meetings

33. Personal and direct

34. Fill in the blank with this word: "___ Bora, wild part of Afghanistan"

35. Digital video file format

36. Type of terrier

38. Short online message

41. Writer Tarbell and others

43. Nissan SUV

46. Unlikely protagonist

49. Fill in the blank with this word: "___ place"

53. Old comic actress ___ Janis

54. Deprives of judgment

55. Tokyo trasher, in a 1956 film

56. Terse bridge bid

57. Fill in the blank with this word: ""___ le roi!""

58. Old character

59. U.S.N. rank below Capt.

61. Pad ___ (noodle dish)

63. Aloud

66. Underwater steerer

PUZZLE 89

ACROSS

1. Soft drink Mr. ___

5. Warty hopper

9. Word that can precede the starts of 17-, 35- and 54-Across and 16-Down

14. Wild Indonesian bovine

15. Fill in the blank with this word: ""Winnie ___ Pu""

16. Trump

17. Wrapper abbr

18. Flying formations

19. Jet similar to a 747

20. Toppled, in a way

22. You're telling me!'

23. The English translation for the french word: sillon

24. Go-aheads

26. Writer LeShan and others

29. With convenience

33. Weary worker's wish

37. Large bra feature

39. Tribe in Manitoba

40. Golden ___ (century plant)

41. Winter coats?

42. Some male dolls

43. Toss

44. Two-base hits: Abbr.

45. Fill in the blank with this word: ""There's ___ chance of that""

46. Nadirs

48. Matador's move

50. Votes against

52. Traditional remedy

57. Worked on Broadway

60. "All aboard!" place

63. The French smoke it

64. Dogpatch possessive

65. Narc's unit

66. B. & O. stop: Abbr.

67. Unusual shoe spec

68. Viennese-born composer ___ von Reznicek

69. Its slogan was once "More bounce to the ounce"

70. Historic Scott

71. Fill in the blank with this word: "Art ___"

DOWN

1. Pampers maker, informally

2. Fill in the blank with this word: "___ alia"

3. Greet respectfully

4. Depth: Prefix

5. Recorded for later viewing

6. Pulitzer-winning author Robert ___ Butler

7. You'll use up 3 vowels playing this word that means toward the side of a ship that's sheltered from the wind

8. Since: Sp.

9. Individuals, so to speak

10. Packs for bikers and hikers

11. Memo abbr.

12. Thought: Prefix

13. Asian, e.g.

21. Winnebago owner, for short

25. Takes some courses?

27. Casa material

28. The English translation for the french word: scalp

30. Songwriter Jacques

31. Filmmaker Riefenstahl

32. Quaint affirmative

33. Obama adviser Emanuel

34. Writing on a French gift tag

35. Norse goddess of fate

36. The English translation for the french word: dÈlinÈer

38. Fill in the blank with this word: "___ Minor"

41. Four hours on the job, perhaps

45. Fill in the blank with this word: ""___ Flux" (Charlize Theron film)"

47. Zany

49. Stood out

51. Kept on the hard drive

53. Voiced one's disapproval

54. Cheat on

55. Ancient Greek tongue: Var.

56. Speed skater ___ Ohno

57. The Bell of ___' (Longfellow poem)

58. Fill in the blank with this word: "Bumper ___"

59. Shortening used in recipes

61. Sorry soul

62. Answer to the riddle "Dressed in summer, naked in winter"

PUZZLE 90

ACROSS

1. German composer Carl

5. Unscramble this word: aetss

10. ___

14. Tree trunk

15. Maine's ___ Bay

16. You'll be the death ___!'

17. Your highness?: Abbr.

18. Letterman airer

19. California tribe that died out in 1916

20. Some yo-yo tricks

22. Much

24. Upticks

25. Writer Santha Rama ___

26. Spacey and namesakes

29. What a D.M.V. issues: Abbr

30. C.I.A. : U.S. :: ___ : Soviet Union

33. Sheep's genus

34. Start of the 18th century

36. Fill in the blank with this word: "___-hoo"

37. Makes a dazzling entrance

41. Santo Domingo greeting

42. Meal, in Milan

43. Three men in ___'

44. Ukraine, e.g., formerly: Abbr.

45. Fill in the blank with this word: "___ la la!'"

46. Begins to transplant

48. Payroll service giant, initially

49. Some early New Yorker cartoons

51. Trading places

54. No tough opponent

58. Hindu Mr.

59. This may work on your block

61. Wild goose

62. Substance from which the universe was created

63. What "Henry" means, literally

64. "Man oh man!"

65. Golf innovator Callaway and bridge maven Culbertson

66. Wide-bodied

67. Variety of chalcedony

DOWN

1. U.K. awards

2. Fill in the blank with this word: ""Let's ___""

3. Unscramble this word: elef

4. Frenetic

5. With 16-Across, way leading to a highway

6. This cavalry weapon was inspired by the Turkish scimitar

7. Exceptional rating

8. Outside: Prefix

9. Soviet comrade

10. Fill in the blank with this word: ""___ want to dance?""

11. Give ___ lip (punch)

12. The English translation for the french word: ‡mha

13. Mariner ___ Ericson

21. Fill in the blank with this word: "___ money"

23. "Alice in Wonderland" sister

26. They're nuts

27. Lesser of two ___

28. Forcefulness

29. Time-share unit

30. This city, imperial capital of Japan before Tokyo, was spared bombing during WWII

31. Leave home

32. Zero-star movies

34. Word to a team

35. Vaccine letters

38. Take on

39. Heavenly gatekeeper, in Portugal

40. Busta Rhymes rhymes

46. Not aching

47. The English translation for the french word: nÙ

48. Green dragon and skunk cabbage

49. Fill in the blank with this word: ""Behold ___ Horse" (William Cooper book)"

50. Old character set

51. make amends for

52. You've got a lot of this, a symbol of bitterness sometimes paired with wormwood

53. Love, honor and ___

55. Fill in the blank with this word: "Dramatist Lope de ___"

56. M.I.T. grad: Abbr.

57. Officially listed: Abbr.

60. Unscramble this word: bur

PUZZLE 91

ACROSS

1. Indian ox

5. What ___ told you ...?'

8. Stopper from a tub?

13. Trix alternative?

14. Fill in the blank with this word: "A ___ technicality"

15. One of Robert Browning's dramatic poems concerned this little mill-girl who "Passes" by

16. Word from the crib

17. What schools have

19. Supply-and-demand subj.

21. Line score letters

22. Fill in the blank with this word: "Eugene ___, hero of "Look Homeward, Angel""

23. Off-road two-wheeler

26. Thing that can bring you down?

28. Verizon FiOS, e.g., for short

29. The English translation for the french word: PME

30. Soldier

31. So's ___ old man!'

32. Rock pioneer ___ Eddy

34. Caught

37. About 4 million Americans, religiously

41. Requests from regulars

42. Some Prado works

44. Rock-___, classic jukebox

47. Fill in the blank with this word: ""Ol' Rockin' ___" (bin-mate of the 1957 album

48. French vineyard

50. Vitamin C source

51. Gathered skirt

53. Sang

55. Fill in the blank with this word: ""If ___ suggest ...""

56. When repeated, a Thor Heyerdahl title

58. Takes root

59. Fill in the blank with this word: "BELL PEPPER ___ BRUSH FIRE"

62. Wayne W. ___, author of "Your Erroneous Zones"

65. Twit

66. Western Indians

67. Yield, as interest

68. Spanish counterparts of mlles.

69. Yellow ___

70. John ___-Davies of the "Lord of the Rings" trilogy

DOWN

1. What a H.S. dropout may get

2. Santa ___, Calif.

3. Like a good golf score

4. Take countermeasures

5. Weapon for Iraqi insurgents: Abbr.

6. Fill in the blank with this word: ""___ Jacques" (children's song)"

7. Fill in the blank with this word: "Architect ___ Ming Pei"

8. Mount ___, active Philippine volcano

9. "The Road" star Mortensen

10. This place is buzzing

11. Fill in the blank with this word: "2001 Economics Nobelist Michael ___"

12. Wine expert, maybe

14. The English translation for the french word: blanche

18. Governmental guarantee

20. Haunted

23. Fill in the blank with this word: ""___ not!""

24. Fill in the blank with this word: ""Which Way ___?" (1977 film)"

25. Nairobi native

27. Transportation charges

30. German chocolate brand

33. Opposite of alt

35. Where the Azores are: Abbr.

36. Fill in the blank with this word: "1957 Physics Nobelist Tsung-___ Lee"

38. Wrap for a nursery plant

39. Kid's taunt

40. Marquis de ___

43. Non sibi ___ patriae' (Navy motto)

44. Writers for old literary magazines

45. someone who has a limp and walks with a hobbling gait

46. Where Noah landed

49. Hearts

52. Lake bordered by Malawi, Mozambique and Tanzania

53. The English translation for the french word: escorte

54. This beverage that can be "sweet" or "hard" is made from freshly pressed deciduous fruits

57. Villainous group in "Get Smart"

60. Zorba imparts the beauty of the Greek language, but some lessons are tricky, like the one teaching that "ne" means this

61. Fill in the blank with this word: ""...___ thousand times...""

63. Trick ending

64. Workers with 64-Downs, for short

PUZZLE 92

ACROSS

1. Yield, as interest

5. Israeli seaport

10. Fill in the blank with this word: ""___ Dead?" (Mark Twain play)"

14. Time: Ger.

15. Rub together

16. Without women

17. Oscar Wilde poem "The Garden of ___"

18. New York City tour provider

19. Wash

20. Walloped, quickly

21. Tickler of the ivories

23. Declares

25. Rhone feeder

26. Kind of cracker

28. Fill in the blank with this word: ""Coffee, ___ Me?""

30. Disprove

31. Turn back

32. Where Charlie "may ride forever," in song

35. Sensory appendage

36. Pattern for a forensic scientist

37. On March 15, 2003 the W.H.O issued a rare emergency travel advisory in response to this outbreak

38. Michigan's ___ Canals

39. Museo del ___

40. All, in stage directions

41. 1986 self-titled album whose cover was Andy Warhol's last work

42. Pre-Russia intl. economic coalition

43. World Series finale

45. Rattletraps

46. Vessels

49. Writer Fleming

52. Fill in the blank with this word: "End ___"

53. Smock

54. Top-___ (leading)

55. Suffix with diet

56. Pa Clampett player on TV

57. George Sand's "___ et lui"

58. Neural network

59. Fill in the blank with this word: "___ prayer"

60. Ron Howard media satire

DOWN

1. O.T. book before Daniel

2. Old magazine ___ Digest

3. 1993 TV western starring Kenny Rogers and Travis Tritt

4. Windows options

5. Selfishness

6. The English translation for the french word: piÉce rapportÈe

7. Stretched out

8. It will come ___ surprise...'

9. Classic Steinbeck story

10. Frequent ferry rider

11. Doesn't go

12. Fill in the blank with this word: ""___ got a girl for you!""

13. Elbe tributary

21. Warm-up for the college-bound

22. Scientology's ___ Hubbard

24. Glove

26. Assns. and orgs.

27. Island in French Polynesia

28. Take on

29. What has made some people miss the mark?

31. India's ___ Jahan

32. Discovery Channel survival show

33. Fill in the blank with this word: "Family ___"

34. Fill in the blank with this word: "Alumni ___: Abbr."

36. Has a lock on?

37. Mr. ___ of "Peter Pan"

39. Warm-up

40. She-bears, south of the border

41. Punish with an arbitrary penalty

42. Fill in the blank with this word: "___ Green, Scottish town famous for runaway weddings"

43. P. C. Wren novel "Beau ___"

44. Succeed in a big way

45. Sculptor Oldenburg

46. Sorry soul

47. Rubaiyat' rhyme scheme

48. Brit's teapot cover

50. Tuna ___

51. Prior: Abbr

54. Three-way joint

PUZZLE 93

[Crossword grid]

ACROSS

1. Obi accessory

5. Year Queen Victoria died

9. Salad tidbit

13. Picasso's muse Dora ___

14. Pas ___ (dance solo)

15. Wore

16. Fur Elise, for one

18. Slalom curves

19. Fill in the blank with this word: "___ Good Feelings"

20. Like a pageant winner

22. Fill in the blank with this word: "Dry ___"

23. Sitter's headache

27. Hall-of-Fame basketball coach Hank

28. Fill in the blank with this word: "Farmer's ___"

29. Scot's exclamation

30. What directors sit on: Abbr.

31. The English translation for the french word: zÈzayer

33. Part of U.S.N.A.: Abbr.

35. Deadlock

37. The Super Bowl, to a football fan

41. Fill in the blank with this word: ""Say that thou ___ forsake me ...": Shak."

44. Lew Wallace's "Ben-___"

45. Quod ___ faciendum

49. French soul

50. Foot part

53. Fill in the blank with this word: "___ Zeppelin"

55. Visitors to the Enterprise

56. Empathic remark, Bard-style?

59. When repeated, a snicker

60. The English translation for the french word: logique

61. Washington-to-Boston speedster

63. "Mr. Mom" co-star and others

64. Nod off

67. Fill in the blank with this word: ""Bonne ___!" (French cry on January 1)"

68. Wide-mouthed pitcher

69. You and who ___?' (fighting words)

70. Fill in the blank with this word: "___ Mary's (L.A. college)"

71. Tiny time unit: Abbr.

72. Made a tax valuation: Abbr.

DOWN

1. Risk

2. Uhuru Park locale

3. Knocking noises

4. Yes-___ question

5. Works for an ed.

6. Wharton grad's aspiration, maybe

7. Prefix with vitamin

8. Repetitively named Philippine province

9. the seventh month of the Islamic calendar

10. Hither and yon

11. Thingums

12. Fill in the blank with this word: ""It ___; be not afraid" (words of Jesus): 2 wds."

15. Walloped

17. Fill in the blank with this word: ""The Bells ___ Mary's""

21. Truman's nuclear agcy.

24. Balance parts

25. Go ___ some length

26. Taoism founder

Lao-___

32. U.K. heads

34. One ___ time: 2 wds.

36. Schubert's '___ Maria'

38. put into service

39. Fill in the blank with this word: "Faulkner's femme fatale ___ Varner"

40. Riley's "___ Went Mad"

41. Fill in the blank with this word: "1957 Physics Nobelist Tsung-___ Lee"

42. Nonhuman part of a cyborg

43. Renders harmless, in a way

46. Fixes, as a pump

47. Dieted

48. Like the bone in a loin steak

51. Robertson of CNN

52. Fill in the blank with this word: "___-Dazs"

54. Strands of biology

57. Snowy ___

58. Works, as a field

62. See 26-Across

63. What a floozy might show off

65. Three-way joint

66. Tolkien used this Old English word for "monster" as a synonym for "goblin"

PUZZLE 94

ACROSS

1. Fill in the blank with this word: "___ signum (here is the proof)"

5. Textbook market shorthand

9. When doubled, a former National Zoo panda

14. Youngest player to join the 500-homer club

15. Provincial capital SW of Beijing

16. Pirouetting, perhaps

17. Trillion: Prefix

18. Midget car-racing org.

19. Had as a base

20. Woeful

21. Mini?

23. "Put me down as a maybe"

25. Unit of hope?

26. Mauna ___ volcano

27. Yucat

30. Usher's offering

33. Pour ___ troubled waters

35. Poll amts.

36. Substance from which the universe was created

37. <--- Plastered

40. Venerable

41. The English translation for the french word: ululer

42. Year the first Rose Bowl was played

43. Writer ___ Louise Huxtable

44. Vaulted

45. Fill in the blank with this word: ""___ note to follow ..."

46. St.-___-l'

47. The English translation for the french word: síapercevoir

50. Round cut

56. Gosh, British-style

57. Letter-shaped fastener

58. This small rodent whose name rhymes with mole is closely related to the lemming

59. The Ponte Vecchio crosses it

60. Sun: Prefix

61. Sci. course

62. Fill in the blank with this word: "Fannie ___ (securities)"

63. David Bowie single with the lyric "If we can sparkle he may land tonight"

64. "Zuckerman Unbound" novelist

65. ...

DOWN

1. Dines at home

2. You should whip this ingredient before you top your Chantilly potatoes with it

3. Mountain chain

4. Soprano Christiane ___-Pierre

5. Rejoice

6. Actress Virna

7. Fill in the blank with this word: "Den ___, Nederland"

8. This unit of measure also means to move by small degrees

9. "What a pity!"

10. Winding

11. Who's there?' answer

12. Koh-i-___ diamond

13. Squire

21. Quaint contraction

22. Tom ___, 1962 A.L. Rookie of the Year

24. Sticky

27. Fill in the blank with this word: "___ point (only so far)"

28. The English translation for the french word: exclusivitÈ

29. Musical notes

30. Steve McQueen's ex-wife and co-star in "The Getaway"

31. Renaissance artist Guido ___

32. Early 26th-century year

33. Workplace watchdog, for short

34. Fill in the blank with this word: ""If ___ You" (#1 Alabama song)"

36. Some child-care center sites, for short

38. Col. Potter on "M*A*S*H," to pals

39. First black N.F.L. Hall-of-Famer ___ Tunnell

44. Fill in the blank with this word: "English author Edward Bulwer-___"

46. Salsa singer Cruz

47. Fill in the blank with this word: ""A merry heart ___ good like a medicine": Proverbs"

48. Blakley of 'Nashville'

49. Texas has one, in song

50. Quarters

51. Yeah, sure!'

52. Fill in the blank with this word: ""Run ___ Run" (1998 film)"

53. Fill in the blank with this word: "___ Kreuger, the Match King"

54. Fill in the blank with this word: ""This will ___ further""

55. About 40 degrees, for N.Y.C.

59. Quantity: Abbr.

PUZZLE 95

ACROSS

1. The English translation for the french word: haleter

5. Workplace for Reps. and Dems.

9. CCCXXVI doubled

14. Wells's oppressed race

15. Start of a 1940's-60's world leader's name

16. Fill in the blank with this word: "Duane ___ (New York City pharmacy chain)"

17. The English translation for the french word: samaritain

19. Urging from Santa

20. Actor Alastair

21. Wrath

22. When the kids are out

24. Stickball relative

26. Fill in the blank with this word: ""And When ___," 1969 Blood, Sweat & Tears hit"

27. U.S. trading partner, formerly

28. Neighbor of Syr.

29. Orator's skill: Abbr.

33. Hair net

36. Weigh station sight

37. Plaintiff

38. Sew shut, as a falcon's eyes

39. Bad lighting?

40. Something good

41. Renaissance artist Guido ___

42. Fill in the blank with this word: ""Laborare est ___ "

("to work is to pray")"

43. Like corduroy

44. Fill in the blank with this word: "___ of Attalos (Greek museum site)"

45. Wedge-shaped inlet

46. Work unit

47. What an A is not

49. Like some coifs

53. Common allergen

56. Schubert's 'The ___-King'

57. Fill in the blank with this word: "___ show"

58. Indian ___

59. Opposite of renown

62. Fill in the blank with this word: ""Laborare est ___ "

"Europe's Gorge of the ___"

63. Fill in the blank with this word: ""Your ___""

64. Fill in the blank with this word: "___ cat"

65. They're well-connected

66. Au ___

67. Fill in the blank with this word: ""Mi casa ___ casa""

DOWN

1. Painting surface

2. Actor Delon

3. W.W. I battle site

4. Fill in the blank with this word: "___ mater"

5. Sarkozy's presidential predecessor

6. Stravinsky's "___ for Wind Instruments"

7. Shelter grp.

8. Federally guaranteed security

9. Descent, as of an airplane

10. Prominently featured

11. Fill in the blank with this word: "___-da (pretentious)"

12. Fill in the blank with this word: "Alter ___ (exact duplicate)"

13. Fill in the blank with this word: ""Able was ___...""

18. Used a kitchen utensil

23. Madison Ave. trade

25. Wind instrument?

28. Arrowsmith's wife

30. Toss

31. Unusual shoe spec

32. Like much folk music: Abbr.

33. Ukr. and Lat., once

34. The English translation for the french word: Tanguy

35. Wine: Prefix

36. "For an avid philatelist like me, sorting envelopes is thrilling - I might spot a ___!"

39. Fill in the blank with this word: ""___ ask ...""

43. in a wry manner

46. Wage ___

48. What road hogs hog

49. Pope John X's successor

50. Wreckage

51. Opera singer Simon ___

52. Fill in the blank with this word: ""___ want to dance?""

53. ___ deck

54. Very light brown

55. Fill in the blank with this word: ""Go, ___!""

60. Fill in the blank with this word: ""...___ thousand times...""

61. Tommy ___, Olympic skiing gold medalist

PUZZLE 96

ACROSS

1. Two of fifty?

5. In South Africa 100 cents equals R1, R standing for this

9. U.S.A.F. rank

14. Sci-fi princess

15. Fill in the blank with this word: "Catch ___"

16. small salmon of northern Pacific coasts and the Great Lakes

17. Chargers' action

19. Spotted ___

20. They may deliver anesthesia

21. Native-born Israelis

22. Manhattan-based fashion co.

23. Object

24. What's right in front of U

25. The yellow & white flower seen here; Donald might "duck" out to get some as a gift

28. an evil spell

30. This is ___'

31. "Compromise is the best and cheapest ___" (saying attributed to Robert Louis Stevenson)

34. Village Voice theater award

37. Having depth

39. Snap

40. New York's ___ Island

41. Fill in the blank with this word: ""You go not till ___ you up a glass": Hamlet"

42. The matador's opponent

44. Universal Studios record label

45. It's not automatic

47. Fill in the blank with this word: ""A little ___ do ya" (1950s-'60s slogan)"

49. Winter hrs. in Bermuda

51. Work without ___

52. Fill in the blank with this word: "Dark ___"

54. Sport in which competitors don't want breaks

56. Books of psalms

60. The English translation for the french word: canal

61. Teaching relationships

62. Fill in the blank with this word: "___ Kristen of "Ryan's Hope""

63. Mae West role

64. Porter ___, former C.I.A. director

65. Start of an opinion

66. Old-fashioned letter opener

67. Go-aheads

DOWN

1. Wells's oppressed race

2. George Manville ___, English adventure writer

3. They take the bait

4. Some lunches

5. You can see an enormous clock on Rue du Gros-Horloge in this city where time ran out for Joan of Arc

6. Fill in the blank with this word: ""12 ___ Men," 58-/46-Across movie (1957)"

7. Unspecified degrees

8. On the ___

9. Toast to one's health

10. Place for playthings

11. Round, red firecracker

12. Chaplin and others

13. Unscramble this word: stela

18. Fill in the blank with this word: "___ halide"

21. Less adorned

23. Italy's longest river

25. Seventh-century date

26. Landers and others

27. Digestive system parts with recycling?

28. Young's 'Father Knows Best' co-star

29. Tiny particle: Abbr.

32. Inexorable process

33. Stout detective Nero ___

35. With: Abbr.

36. Alike: Fr.

38. Woe while getting clean

43. Newbery-winning author Scott ___

46. "The Furys" novelist James

48. Fill in the blank with this word: ""A kingdom for ___": "Henry V""

49. Fill in the blank with this word: "___ art (text graphics)"

50. Viking poet

52. Up and about

53. Scottish Highlanders

55. Suffix for a collection

56. Woolly-coated dog

57. They're inflatable

58. Adjust, as a clock

59. Sound of a leak

61. Y. A. Tittle scores

PUZZLE 97

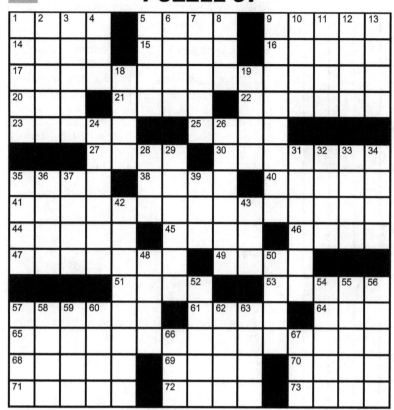

ACROSS

1. This member of the parsley family with a 4-letter name is a main flavoring agent in pickles

5. Stretch ___

9. The English translation for the french word: Delhi

14. Fill in the blank with this word: "___ in a blue moon"

15. Grp. organizing '60s sit-ins

16. Whopper topper

17. Ones concerned with sustainable design

20. Where to see a mummy: Abbr.

21. Willow variety

22. Twisty items

23. Down Under soldier

25. Start of a pirate's chant

27. Manhattan-based fashion co.

30. Blood pigment

35. Voodoo

38. Unscramble this word: liar

40. Fort ___, Md.

41. Create a whole new set of problems

44. O.K.s, in Toledo

45. Granny ___

46. To remove, as from office

47. Pizza topping choice

49. Underwater growth

51. In ___ of (replacing)

53. Fill in the blank with this word: ""The stage ___""

57. Offer advice from around a card table

61. Pitcher Hideo ___

64. Verb type: Abbr.

65. Sister of Joseph II

68. Have ___ of mystery

69. Things that are commonplace are a this "a dozen"

70. Warp

71. Simon Wiesenthal's quarry

72. Ja and da

73. Words of disrespect

DOWN

1. You'd better believe it

2. Ski jump downslope

3. Hate

4. Washington and ___ University

5. Would-be J.D.'s hurdle

6. Words often before a colon

7. Fill in the blank with this word: ""Little Nemo" cartoonist Winsor ___"

8. Scot's exclamation

9. Semicolon?

10. D.O.E. part: Abbr.

11. Vermin

12. Serious trouble

13. Without restraint

18. Small cut

19. Fill in the blank with this word: ""___ Dead?" (Mark Twain play)"

24. Wounded by Cupid's arrow, Venus fell in love with this handsome guy at 1st sight

26. Hey, check that out!'

28. Wayne LaPierre's org

29. Fill in the blank with this word: "___ Stadium, opened in 1923"

31. Fill in the blank with this word: "___ Fables"

32. Popular fragrance

33. Wedding exchange

34. Word to waiters

35. Spanish ___

36. Suffix with my-

37. Proper name in Masses

39. Insurance card info

42. Books of psalms

43. Fill in the blank with this word: "___ 1 (Me.-to-Fla. highway)"

48. Great Sphinx locale

50. Literary ___

52. The English translation for the french word: dÈtacher

54. Fill in the blank with this word: "Alaska's ___ National Historical Park"

55. The sculptures "Rigoletto" and "La Tosca," e.g.

56. Highland pants

57. Essential company figure

58. Suffix for a collection

59. Arg. neighbor

60. Peer group?

62. Spy Mata ___

63. Fill in the blank with this word: "Architect Ludwig ___ van der Rohe"

66. -

67. Winding road shape

PUZZLE 98

ACROSS

1. Hip

5. Forest ___

10. Under a quilt, say

14. Spy Mata ___

15. Fill in the blank with this word: ""Beat ___ to ..."

16. Suffix with utter

17. Chekhov play

19. Side dishes: Abbr.

20. Where to find baked blackbirds

21. Judo maneuvers

23. Super ___ (video game name)

24. Leadership org. opposed to the G.O.P.

26. When it's broken, that's good

27. Leave a neighborhood gym?

33. Weary worker's wish

36. Jazz standard '___ Me'

37. Fill in the blank with this word: "___ show"

38. Give ___ on the shoulder

39. B and O figures: Abbr.

40. Transcontinental bridge, e.g.: Abbr.

41. Texas N.B.A.'er

42. Wheat bundle

43. When to say "Feliz A

44. A trusting person may be led down it

47. Mr. Rogers

48. Fill in the blank with this word: "Country music's ___ Brown Band"

49. Year that Dionysius of Halicarnassus is believed to have died

52. Pilot

57. Sound before a blessing

59. Golf's ___ Aoki

60. Blue-green gem

62. The English translation for the french word: tache

63. Fill in the blank with this word: ""Is ___ joke?""

64. South Asians speak it

65. Film director Petri

66. Thwart in court

67. Fill in the blank with this word: "___-fry"

DOWN

1. Wanna play?'

2. Department of central France

3. Workout spots, for some

4. Squeal in pain

5. Fill in the blank with this word: ""___ you heard?""

6. Sam Adams Rebel ___

7. Pop singer ___ Del Rey

8. Long dist.

9. They clean locks

10. Prance about

11. "Currently serving" military designation

12. Pour

13. Boot camp affirmative

18. Where to get down

22. Fill in the blank with this word: "___ riot (very funny skit)"

25. Actress Phyllis of "I Was a Teenage Frankenstein"

27. The English translation for the french word: PIB

28. Fill in the blank with this word: "Caput ___ syndrome (arm problem)"

29. Reveals one's feelings: 2 wds.

30. Landing spot for 74-Down

31. Old Testament book

32. Start of a pirate's chant

33. Wheelchair access

34. essential oil or perfume obtained from flowers

35. Peaceful race in "Avatar"

39. Salty hail

40. Foot part

42. Some low-income housing, for short

43. ___ Sketch (classic drawing toy)

45. The last novel featuring him was "Stopover: Tokyo"

46. Pre-GPS guide from a travel org

49. When someone starts to get riled, he's told to "keep" this garment "on"

50. Actress Beulah

51. Heart of France

52. Unscramble this word: dhie

53. Dusseldorf donkey

54. Legal scholar Guinier

55. Fill in the blank with this word: "Cutty ___ whisky"

56. Ravel's "Gaspard de la ___"

58. Vineyards of high quality

61. Japanese prime minister Taro ___

PUZZLE 99

ACROSS

1. The English translation for the french word: ruinure

6. Atlantic food fish

10. Twerp

14. Vermont's ___ Mountain Resort

15. Shakespearean king

16. Actress Skye

17. Fill in the blank with this word: ""___ a good-night!""

18. Directories

20. Westminster Show org.

21. Time for eggnog

23. Fill in the blank with this word: "___ the heart of"

24. Just let ___'

25. Position in a rock band

27. Happy ex-Mayor of New York?

31. Z-car brand

32. Middle of a run?

33. I don't mean to ___ ...'

36. Supermodel Wek

37. Cornerback Sanders

39. "Trinity" novelist

40. Worker in a garden

41. Pop singer Manfred ___

42. Pie chart section, perhaps

44. Icon

46. a member of a Bantu speaking people living in Rwanda and Burundi

49. Animal with striped legs

50. Missed by ___ (was way off)

51. Zodiacal delineation

52. Fill in the blank with this word: ""You've ___ Mail""

55. Totally involved with

58. Thomas who wrote "Common Sense"

60. Fill in the blank with this word: "___-mutuel"

61. With 33-Across, anagrams and puns (or parts hidden in 17-, 24-, 44- and 51-Across)

62. Sharp-___

63. Way: Abbr.

64. Hollywood's Roberts and others

65. Olive ___

DOWN

1. Fill in the blank with this word: "___ one"

2. Relenting assent

3. Sound system brand

4. Year that Eric the Red was born, traditionally

5. Pre-Easter time

6. Zigzag, in a way

7. Yield

8. The English translation for the french word: RAU

9. Grand ___, Nova Scotia

10. Shin bones

11. Tempter

12. Stamping need

13. What the doctor ordered

19. Replay speed

22. Some Lake Victoria viewers

24. Fill in the blank with this word: ""Now ___ you...""

25. Fill in the blank with this word: "___ sabe"

26. Oatcakes popular in Scotland

27. Just ___ (very little)

28. The English translation for the french word: ballot

29. Way from Syracuse, N.Y., to Harrisburg, Pa.

30. Traditional Shrovetide dish served with caviar and sour cream

33. Fill in the blank with this word: "___ and proper"

34. Vex

35. Fill in the blank with this word: "Battle of the ___, opened on 10/16/1914"

38. Wyoming Senator Mike

39. Western Athletic Conference sch.

41. Some church music

43. Rough and reddy?

44. Sermon site

45. Plumbs the depths

46. Workers with wings

47. Orlando's ___ Arena

48. Levels

51. Publisher's concern: Abbr.

52. This prefix turns 1 byte of data into about a billion

53. Perfect report card spoiler

54. Fill in the blank with this word: ""Bill & ___ Excellent Adventure""

56. Hockey's Krupp

57. Taro dish

59. Worker in the TV biz

PUZZLE 100

ACROSS

1. Fill in the blank with this word: "Ad ___ per aspera (Kansas' motto)"

6. Sugar amts.

10. Fill in the blank with this word: ""___ she blows!""

14. Zones

15. Tombstone name

16. Spear

17. Jungle swinger

20. Woe ___!'

21. Univ.

22. Set one's sights

23. Bottled spirit?: Var.

25. Unto us ___ is given': Isaiah

27. Young women's grp

30. Liquid ___

32. Wriggler

36. Fill in the blank with this word: ""If ___ Would Leave You""

38. Fill in the blank with this word: "___ out (relax, like, totally)"

39. Fill in the blank with this word: ""And When ___," 1969 Blood, Sweat & Tears hit"

40. It often has islands

44. Sand ___

45. Pitcher Robb ___

46. Israel's Barak and Olmert

47. Word often prefixed with kilo-

48. Acquire again

51. Give __ break!'

52. Youngest player to join the 500-homer club

54. Three-time N.H.L. All-Star Kovalchuk

56. Fill in the blank with this word: "Agatha Christie's "There Is ___"

59. Fill in the blank with this word: "Alumni ___: Abbr."

61. 1974 Sutherland/Gould spoof

65. Felt bitter anguish

68. Finish this popular saying: "As you sow so shall you_____."

69. Heyerdahl's second papyrus boat

70. Part of a sentence, in linguistics

71. Duck: Ger.

72. Some hosp. workers

73. Went after

DOWN

1. They follow so

2. Principal river of Armenia

3. One of the Munsters

4. Vegged out

5. Opposite of SSW

6. Yellowstone range

7. Wry comic Mort

8. Hierarchs

9. Fill in the blank with this word: "Day ___"

10. Connect to

11. Soccer star Mia

12. Zoological wings

13. Tear violently

18. Spinoff series with two spinoffs of its own

19. Kung ___ chicken

24. Set in "Die Fledermaus"

26. USMC rank

27. Have a spot ___

28. The English translation for the french word: ovaire

29. Plopped down again

31. Fill in the blank with this word: ""___ could have told you that!""

33. Widespread loathing

34. Fill in the blank with this word: "___ Oro"

35. Wise guys?

37. Signature Muhammad Ali ploy

41. Obi accessory

42. Can't wait to have

43. Unsubstantial

49. Some Japanese-Americans: Var.

50. Wrist-elbow connector

53. TV pooch

55. Fill in the blank with this word: ""___ World Turns""

56. Fill in the blank with this word: "Europe's Gorge of the ___"

57. Kind of list

58. Fill in the blank with this word: ""I earn that ___": "As You Like It""

60. Tibia

62. O'Neill's "A Touch of the ___"

63. Where the Gila joins the Colorado

64. Put back

66. Schubert's 'The ___-King'

67. They have Xings

SOLUTIONS

PUZZLE SOLUTION 1

```
T O N E . . S A H L . . T A L U S
O M A N . . T H E E . . E L I S A
O N N O . . O O R T . . C A F F E
K I C K U P Y O U R H E E L S . .
H A Y I N G . N P I N . . . . . .
. . . F O U R . C O N M A N . . .
O B O L I . G Y R A . . A A B A .
F R A C T A L . A N S P A C H . .
I N S T . M I T T . U S R D A . .
T O T O I V . A S A D . . . . . .
. . . N E A P . S O B E R S . . .
W H A T S T H E B I G D E A L . .
R I C K I . O M I T . . O N T O .
A T H O S . S A T I . . O I S E .
P S S S T . T N T S . . R E O S .
```

PUZZLE SOLUTION 2

```
N E D D A . E W A . S T G E R
U L E E S . Y I N . W H I F F
N I E L S . E L A . O E U F S
C O M E D Y C E N T R A L . .
I T I S . I H E A R D . I I S
O S T . U P A . B I S T A T E
. . P E E R S . . . I N I S .
P A I D A S T E E P P R I C E
I T S A . . X F I L E . . . .
K R I S H N A . F A Y . A I N
E A T . A I R G U N . D E N E
. . T H A T I S S O G R O S S
S P R I G . D P I . E L L I S
A R U L E . N O V . E A U D E
T I E O N . I T E . R O S E S
```

PUZZLE SOLUTION 3

```
M Y W A Y . W R E S T . A D O
S E E N A . H E T T Y . B O N
N O B E L P E A C E P R I Z E
. . M I E N . S P E E D E E .
C U T I E P I E . E V E R Y .
A T I C . I N I . A R S E .
R E D . S T A G E M O M . . .
P S Y C H O T H E R A P I S T
. . L U N E T T E S . B E E .
D O L E . S S H . S E A R .
O L A V S . S W E E T G U M .
A D D E N D A . E L M A . . .
B E D R O O M S L I P P E R S
L S I . R H O M B . T E P E E
E T E . E A R L Y . Y S A Y E
```

PUZZLE SOLUTION 4

```
G A M E D . G R I G . I S N O
O C U L I . R O T O . D T E N
T H A I S . I N I T . N A Z I
M E N S C H F A S H I O N . .
E R G . T O F . N I M . D A E
. . N U D I S T C O L O N Y .
L A L O . G T O . F O U N D .
A L O T . E H U D S . P T U I
T C O A T . L I N . E S M E .
E A S T E R I S L A N D . . .
N N E . N E N . A K U . R K O
. B L A N K E T Y B L A N K .
C H A I . F I D O . I M B U E
S A L T . R E D R . A N A R M
A L L H . O R A Y . N O T L O
```

PUZZLE SOLUTION 5

O	P	I	U	M	■	R	O	T	C	■	A	R	A	B
D	E	S	N	A	■	E	R	O	O	■	N	I	T	S
D	R	A	F	T	P	I	C	K	S	■	N	I	L	E
D	O	R	E	M	I	■	S	Y	M	P	O	S	I	A
■	■	■	D	A	Z	S	■	O	I	L	Y	■	■	■
B	O	T	■	N	A	P	E	■	C	A	I	M	A	N
A	N	E	■	■	Z	I	G	S	■	C	N	O	T	E
L	E	N	O	Z	Z	E	D	I	F	I	G	A	R	O
S	I	L	V	A	■	R	O	L	O	■	■	T	A	N
A	L	B	E	I	T	■	N	K	G	B	■	S	S	S
■	■	■	R	R	U	N	■	S	E	R	E	■	■	■
I	C	E	B	E	R	G	S	■	Y	O	G	U	R	T
D	O	D	I	■	N	A	M	E	S	N	A	M	E	S
O	S	E	T	■	I	I	I	I	■	C	L	A	N	G
L	Y	L	E	■	P	O	T	S	■	O	S	S	E	T

PUZZLE SOLUTION 6

A	M	A	J	■	M	A	I	D	S	■	H	R	S	■
C	A	F	E	S	■	O	R	S	E	A	■	E	I	N
H	U	F	F	A	N	D	P	U	F	F	■	L	G	A
■	■	I	F	N	I	■	S	L	E	E	T	I	E	R
S	N	L	■	A	G	N	■	I	N	A	H	O	L	E
D	O	I	■	A	H	O	Y	■	D	R	E	■	■	■
N	I	A	S	■	T	H	A	W	■	E	M	C	E	E
A	S	T	U	D	Y	I	N	S	C	A	R	L	E	T
H	Y	E	N	A	■	T	N	T	S	■	S	A	N	A
■	■	■	S	N	I	■	I	O	T	A	■	M	S	G
A	C	H	E	F	O	R	■	N	O	M	■	B	Y	E
T	R	I	T	O	N	A	L	■	R	O	J	A	■	■
A	I	T	■	R	I	D	D	L	E	C	A	K	E	S
L	S	I	■	T	Z	A	R	A	■	O	D	E	O	N
L	S	T	■	H	E	R	S	H	■	■	A	S	S	E

PUZZLE SOLUTION 7

C	A	H	N	■	M	E	S	A	S	■	O	D	A	S
A	D	A	M	■	E	A	R	T	H	■	R	I	N	A
R	A	D	I	O	D	R	A	M	A	■	O	R	G	S
L	M	N	■	L	I	S	S	O	M	■	M	E	E	K
A	T	O	S	S	A	■	■	P	S	E	C	■	■	■
■	■	S	E	L	F	P	R	O	M	O	T	E	R	■
O	G	D	E	N	■	L	T	C	O	L	■	H	E	E
T	H	A	E	■	D	E	E	P	S	■	G	I	R	L
I	I	N	■	Q	B	E	R	T	■	P	A	T	S	Y
C	A	M	E	A	C	R	O	S	S	E	D	■	■	■
■	■	A	C	T	O	■	■	A	R	S	O	N	S	■
E	A	R	H	■	O	T	E	L	L	O	■	P	A	T
O	L	I	O	■	P	E	N	M	A	N	S	H	I	P
N	E	N	E	■	E	L	A	N	D	■	M	I	L	A
S	N	O	R	■	R	A	M	O	S	■	E	R	S	T

PUZZLE SOLUTION 8

I	D	A	H	O	■	C	O	O	N	■	S	E	A	M
P	I	P	E	S	■	O	T	O	E	■	E	R	M	A
S	U	P	P	L	E	M	E	N	T	A	T	I	O	N
■	■	■	E	C	O	L	E	■	B	E	A	M	E	D
D	B	A	■	O	T	R	A	■	R	E	B	B	E	■
A	L	L	R	A	■	H	U	L	A	O	■	M	A	L
N	A	T	U	R	E	■	B	I	L	L	G	■	■	■
■	H	O	T	C	R	O	S	S	B	O	N	E	S	■
■	■	S	H	I	R	R	■	A	G	A	S	S	I	■
S	F	C	■	I	S	A	A	K	■	Y	R	O	S	E
K	I	R	O	V	■	D	W	A	N	■	■	P	S	S
I	L	O	N	A	S	■	■	H	E	A	T	H	■	■
P	E	O	P	L	E	S	F	U	N	E	R	A	L	S
I	N	N	O	■	Q	U	I	N	■	R	E	G	A	L
T	E	S	T	■	S	E	G	A	■	O	K	I	E	S

PUZZLE SOLUTION 9

Y	E	A	S	T		S	E	D	A		L	S	A	T
O	C	H	O	A		P	N	I	N		E	E	R	O
D	R	I	L	L	T	E	A	M	S		G	E	E	R
S	U	S	A		S	A	M	S	A		A	G	U	T
		C	I	T	R	O		R	E	T	A	P	E	
P	O	S	E	U	R		R	E	A	R	E	R		
S	R	O		S	A	Y	S	T		S	E	L	L	S
H	E	L		E	P	I		H	I	K		A	E	S
C	O	D	E	D		N	A	N	C	I		N	A	E
		I	D	T	A	G	S		I	N	I	D	L	E
O	R	E	G	O	N		S	A	N	E	R			
A	E	R	I		T	H	I	N	G		R	A	H	M
S	T	A	E		R	E	G	I	S	T	E	R	E	D
E	I	N	S		U	R	N	S		I	G	N	I	S
S	E	T	T		M	O	S	E		A	S	O	R	E

PUZZLE SOLUTION 10

S	C	R	A	P	P	A	P	E	R		S	E	L	Z
A	R	E	W	E	A	L	O	N	E		A	N	O	A
T	E	N	N	E	S	S	E	A	N		U	S	M	C
I	T	A	S		S	A	M		A	S	T	A	B	
V	A	T		Y	U	B	A	N		T	E	L	A	R
A	N	O	K	A			I	R	A		A	R	I	
		O	K	E		H	O	L	Y		D	D	E	
R	E	N	A	I	S	S	A	N	C	E	M	A	I	N
A	X	E		T	E	E	M		G	D	P			
D	O	W		O	S	E			A	T	E	O	F	
O	R	S	E	R		K	M	A	R	T		S	R	A
	C	D	I	I	I		A	D	O		A	T	A	R
L	I	E	N		D	E	S	I	D	E	R	A	T	A
O	S	S	E		E	N	U	M	E	R	A	T	E	D
S	T	K	S		M	O	R	E	O	R	L	E	S	S

PUZZLE SOLUTION 11

A	C	C	R	A		A	C	H	Y		P	E	R	I
S	H	O	E	D		F	L	O	E		A	R	U	M
H	O	W	D	O	Y	O	U	D	O		P	E	E	P
Y	I	P		B	O	O	B		M	A	Y			
	R	U	P	E	R	T		B	A	R	R	A	G	E
	N	O	S	E		S	E	N	T	I	E	N	T	
P	A	C	K		W	E	A	R	Y		R	A	H	
A	C	H	E		D	A	N	D	Y		M	O	R	E
T	O	E		S	I	S	S	Y		U	S	S	R	
I	R	R	I	T	A	T	E		H	A	L	O		
O	N	S	T	A	G	E		L	A	B	E	L	S	
		A	B	O		S	I	L	L		B	U	R	
P	A	I	L		N	E	T	T	L	E	S	O	M	E
U	G	L	I		A	M	A	H		S	O	M	M	E
T	A	L	C		L	U	G	E		T	U	B	A	L

PUZZLE SOLUTION 12

O	N	C	E		R	U	B	S			T	B	S	P
P	A	L	M		I	R	A	Q		R	U	L	E	R
T	R	O	U	S	S	E	A	U		O	N	I	C	E
I	R	A		P	E	A		A	S	S	I	S	T	
M	O	C	H	A	S		T	R	I	A	S	S	I	C
A	W	A	Y		F	E	E	L			F	L	U	
		M	E	D	I	A		I	S	S	U	E	R	
	I	N	T	E	R	R	A	C	I	A	L			
B	A	L	S	A	M		G	N	A	R	L			
U	R	L		O	K	A	Y		S	A	K	I		
T	A	F	F	E	T	A	S		T	E	A	S	E	R
	P	A	L	T	E	R		C	A	B		S	E	A
E	A	T	I	N		S	N	O	W	B	O	U	N	D
T	H	E	T	A		T	E	R	N		F	R	E	E
C	O	D	S		S	W	A	Y			T	E	D	S

PUZZLE SOLUTION 13

S	M	O	K	E			I	D	E	S			V	E	L	A
T	I	B	I	A			N	A	S	A			I	R	A	N
A	N	E	N	T			L	I	O	N	I	Z	I	N	G	
B	U	R	G	O	M	A	S	T	E	R				G	O	O
L	E	O			N	E	W			E	R	E			E	L	L
E	T	N	A			T	S	A	R				A	R	I	A
			F	O	R			M	I	S	S	I	O	N		
		V	I	V	I	S	E	C	T	I	O	N		
		S	A	R	A	C	E	N			U	R	L	
S	I	N	E			A	D	A	R			I	O	N	S
I	R	A			P	A	L			D	D	T			R	O	W
E	L	D			A	M	E	T	H	Y	S	T	I	N	E
R	O	A	D	M	O	V	I	E			A	H	E	A	D
R	I	T	E			L	E	E	R			R	A	N	G	E
A	N	E	W			E	L	S	E			S	I	T	E	S

PUZZLE SOLUTION 14

S	N	O	W			P	L	E	B			A	I	S	L	E
T	I	R	O			H	O	M	E			L	I	P	I	D
E	S	A	U			A	B	E	T			P	I	E	C	E
P	I	L	L	A	R			R	A	J	A			E	E	N
				D	I	M	L	Y			I	C	E	D	
S	C	A	N	D	A	L				G	A	R	L	I	C
C	O	P	T			C	A	L	L	S			A	I	N	U
A	M	P			O	M	A	H	A				M	A	R
B	E	E	P			P	A	P	A	W			P	I	N	E
S	T	A	T	U	E			S	P	O	R	T	E	D	
			S	A	R	I			M	A	U	V	E		
A	H	A			G	A	Z	A			Z	A	F	T	I	G
S	A	B	L	E			O	Y	E	Z			A	B	L	E
A	L	L	E	N			O	V	A	L			C	A	L	L
P	L	E	A	T			M	E	R	E			E	R	S	T

PUZZLE SOLUTION 15

A	L	B	S			A	L	T	O	S			L	I	S	P
S	A	R	I			L	E	A	S	T			A	N	T	E
I	D	O	L			B	I	B	L	E			T	S	A	R
F	Y	K	E	S				L	O	N	G	T	E	R	M
		E	X	C	U	S	E			C	H	I	C		
P	U	N			A	R	C	O			H	E	C	T	O	R
A	S	H			D	E	A	F			E	E	R	I	E
G	A	E	L			A	R	C	E	D			S	E	L	F
A	G	A	I	N			O	V	E	R			P	E	E
N	E	R	V	E	D			N	E	V	E			E	R	R
		T	E	A	R			T	R	A	V	E	L		
A	P	E	R	T	U	R	E			S	A	L	A	L
M	I	D	I			D	Y	N	E	S			V	E	D	A
I	S	L	E			G	E	T	T	O			E	N	D	S
D	A	Y	S			E	S	S	A	Y			S	T	Y	E

PUZZLE SOLUTION 16

T	H	I	E	U			P	A	R	C			I	T	G	O
W	I	T	T	S			E	P	H	A			S	Y	L	L
O	E	S	T	E			D	R	U	M			I	P	A	D
A	D	I	E	U	A	D	I	E	U			D	E	S	E
				P	I	L	E			S	P	R	A	Y	S
E	T	R	E			R	E	S	T			N	O	B	E	T
D	R	Y	L	A	W			T	I	N	S				
B	A	N	D	S	A	W			S	F	U	M	A	T	O
			P	R	O	S			O	P	T	S	I	N
C	R	O	R	E			N	C	A	R			N	I	C	K
D	O	M	I	N	I			E	S	T	D				
C	L	E	F			P	A	N	S	Y	Y	O	K	U	M
A	L	A	E			A	L	I	I			K	E	E	P	A
S	E	R	S			D	L	C	S			E	Y	R	I	E
E	R	A	T			S	Y	S	T			S	E	N	S	O

PUZZLE SOLUTION 17

```
L C M S . M S E D . E S T A S
C H A O . N C A A . X T I N A
D A R T B O A R D . C R E O N
T R Y . A P T . A R I A D N E
V W J F G Q . . . E A U . . .
. . A P R I C O T . B B Q S .
R I S H I . C A L E B . L U I
F O O D P R E P A R A T I O N
D U O . E Y E O F . T O N I O
S S N S . A S S I G N S . . .
. T O N . . . . R E S I N S .
H E R O I S M . Q I X . N O O
I F E L L . C O U N T D O W N
L O D G E . M E A D . S N A G
T R I A D . I N D S . L E Y S
```

PUZZLE SOLUTION 18

```
D E P O T . Y V E S . A B O O
A V O E R . O I N K . S E P S
W A S N O . K N E E . A S E S
. . T O P S E C R E T C O D E
L S H . H I D . I A M S O . .
I N A F I X . B F L A T . . .
G O S E E . T I T I . B S S .
N O T E S O F T H E S C A L E
E K E . R A E S . A H S I N .
. J E U L R . O V I E D O . .
A N N A L . . S U A . B E R .
P S I S A N D W E I G H T . .
H I T C . O R A L . E Y E H S
I D A H . G A I A . L U A U S
S E L A . S G T S . Y A M A S
```

PUZZLE SOLUTION 19

```
P P P S . T P E D A . H G T S
R A E S . O R L O P . B O O E
I N T H E H O U S E . E S T D
C A R . L O L L . R N A S E .
E M E . O E O . B I G T I M E
R A L P H . G A G O N . P I T
. H I F . E D D . Y E C H . .
. V I M A N D V I G O R . . .
B R A Z . I C E . C E N . . .
F O G . A N A S S . I D E S T
A M A N I T A . I T S . T O R
. A R E G O . S N E E . H I E
U N I E . D I N G A L I N G S
P I E D . O C U L O . B I N S
N A S A . R U G E R . A C E Y
```

PUZZLE SOLUTION 20

```
C E S T . D O U G . I D I N G
O N E H . I S L A . N O T A R
A S L O O S E A S A G O O S E
C U M U L I . N U B . R O I S
H E A . A N S . P R I M . . .
. . N T H S . I S A Y S O . .
S V E . E I C S . A T E U P .
T I N K E R S H O R T S T O P
I D O N T . H W F E . I S S .
T A L I A S . A T M O . . . .
. T H O R . Y A D . A D A . .
I N S T . I U D . I D I G I T
T H E E A R T H A N D M A R S
A R A R A . H O W E . S I T I
N A S S A . S W E D . O N Y X
```

PUZZLE SOLUTION 21

F	L	A	I	R		S	O	B	S		S	H	O	O
O	U	N	C	E		A	C	A	I		A	E	R	Y
G	R	E	E	N	S	W	A	R	D		F	R	E	E
Y	E	W		T	O	O		B	E	L	A	B	O	R
		C	A	L	F		E	M	I	R				
	S	P	E	C	I	F	I	C	A	T	I	O	N	
D	O	O	D	A	D		N	U	N		R	U	B	
I	L	L	E	R		A	C	E		A	S	I	D	E
G	U	Y			S	K	A		C	R	I	N	G	E
	S	P	I	T	T	I	N	G	I	M	A	G	E	
			C	O	R	N		A	T	O	M			
L	E	G	A	T	E	E		P	E	R		O	T	T
O	D	O	R			A	S	P	I	D	I	S	T	R	A
B	E	A	U		M	I	E	N		A	P	I	U	M
E	N	D	S		S	A	N	G		L	Y	S	E	S

PUZZLE SOLUTION 22

S	L	A	V		F	E	A	T		I	M	B	U	E
N	O	N	E		U	N	D	O		N	A	I	R	A
O	U	T	L	A	N	D	E	R		L	I	N	E	R
T	R	I	A	L	R	U	N		P	A	N	D	A	S
			C	U	E		B	A	W	L				
S	A	L	M	O	N		C	I	G		Y	E	A	H
T	R	O	U	T		O	O	Z	E		L	A	Y	
R	O	O	M	T	E	M	P	E	R	A	T	U	R	E
U	M	P			R	E	S	T		L	E	D	O	N
M	A	Y	A		A	G	E		G	A	L	E	N	A
			M	E	S	A		B	A	R				
U	P	S	I	D	E		M	A	R	M	I	T	E	S
S	O	N	D	E		D	U	S	T	S	T	O	R	M
P	R	I	S	M		A	L	E	E		C	L	I	O
S	E	P	T	A		H	E	R	R		H	U	N	G

PUZZLE SOLUTION 23

L	A	T	T	E		D	R	E	A	D		G	S	A
A	C	O	R	N		E	E	R	I	E		R	A	N
P	H	R	A	S	A	L	V	E	R	B		A	R	T
P	E	E	N		N	E	E		R	U	P	E	E	
		C	U	T	T	L	E	F	I	S	H	E	S	
W	H	E	E	L	I	E		A	R	E	A			
H	A	D	S	T			G	R	A	F	F	I	T	I
I	T	D		G	U	E	S	T			D	E	C	
R	E	A	D	J	U	S	T		C	R	O	A	K	
			Y	U	L	E		U	S	U	A	L	L	Y
S	N	E	A	K	P	R	E	V	I	E	W			
H	O	R	D	E		N	U	T		H	Y	P	O	
U	R	N		B	A	T	T	L	E	F	I	E	L	D
S	I	S		O	P	E	R	A		E	D	G	E	D
H	A	T		X	R	A	Y	S		D	E	G	A	S

PUZZLE SOLUTION 24

E	R	N	S		J	S	U	P		S	A	I	L	S
P	E	A	U		O	H	N	E		C	P	L	U	S
H	A	L	F	P	R	I	C	E		R	I	E	N	S
A	S	A		A	J	A		L	E	A	R	N	S	
		R	S	A		E	R	L	E	S				
B	L	O	A	T		P	L	E	O	N	A	S	M	S
A	Y	E	N		S	U	E	T	Y		W	E	A	N
M	R	S	D		E	L	C	I	D		A	T	M	O
B	I	T	O		E	S	T	E	S		P	H	E	W
I	C	E	M	A	K	E	R	S		P	I	S	T	E
			N	O	T	S	O		S	T	A			
T	E	S	O	R	O		R	A	U		A	V	S	
T	R	A	I	T		F	O	R	H	I	T	L	E	R
O	D	E	S	A		A	R	E	E		H	A	N	A
P	A	S	E	S		A	D	D	L		O	S	A	S

PUZZLE SOLUTION 25

R	S	V	■	L	E	G	T	O	■	H	I	S	T	O
E	M	I	■	C	R	E	A	M	■	D	O	T	E	R
D	I	V	I	D	E	D	C	A	P	I	T	A	L	S
S	L	I	T	■	■	■	T	H	E	R	A	M	■	■
K	L	A	T	S	C	H	■	A	D	J	■	P	I	T
Y	A	N	■	T	U	E	S	■	E	B	B	E	T	S
■	■	K	A	R	S	T	S	■	■	G	D	A	Y	■
■	V	O	W	E	L	S	I	N	O	R	D	E	R	■
T	I	N	A	■	■	E	R	I	C	I	V	■	■	■
D	O	T	I	N	G	■	S	P	A	R	■	R	H	E
P	L	O	■	E	A	P	■	S	N	E	A	K	E	D
■	■	G	E	A	R	E	D	■	■	B	E	Y	E	■
T	W	E	N	T	Y	N	I	N	E	P	A	L	M	S
P	E	N	N	E	■	T	C	E	L	L	■	L	O	S
S	A	Y	E	R	■	A	E	G	I	S	■	Y	M	A

PUZZLE SOLUTION 26

H	A	D	A	C	■	A	T	E	A	■	E	T	T	A
O	C	U	L	I	■	R	A	K	I	■	S	H	A	H
H	E	L	E	N	H	A	Y	E	S	■	P	A	V	O
■	■	C	E	C	U	M	■	■	H	P	O	W	E	R
I	F	I	■	H	A	I	R	R	A	I	S	E	R	S
R	A	N	S	■	D	E	O	■	L	O	R	N	E	■
T	H	E	I	S	M	■	P	U	P	A	■	■	■	■
■	D	A	R	T	M	O	U	T	H	F	A	N	S	■
■	■	R	E	G	T	■	S	S	K	I	N	S	■	■
S	A	T	I	E	■	G	E	T	■	C	H	I	A	■
P	R	I	M	E	R	I	D	I	A	N	■	I	P	O
E	R	R	A	T	A	■	R	A	T	E	L	■	■	■
L	A	I	R	■	I	A	M	A	M	E	R	I	C	A
L	Y	N	E	■	N	O	U	N	■	S	O	S	A	D
S	S	G	T	■	S	N	Y	E	■	T	O	T	A	L

PUZZLE SOLUTION 27

O	C	T	■	F	A	B	■	M	A	P	■	D	N	A
C	H	I	C	A	G	O	■	A	L	I	■	E	E	L
T	I	M	E	C	A	P	S	U	L	E	■	M	O	B
A	M	B	L	E	R	■	O	V	E	R	D	O	■	■
V	E	R	T	■	A	L	L	E	G	I	A	N	C	E
E	R	E	■	E	G	O	■	E	S	P	I	A	L	■
■	■	I	M	A	G	O	■	■	H	A	R	K	■	■
C	R	O	S	S	R	E	F	E	R	E	N	C	E	S
E	A	R	L	■	■	T	R	I	N	E	■	■	■	■
S	N	E	A	K	S	■	I	N	S	■	P	O	P	■
S	I	G	N	A	L	L	I	N	G	■	P	I	P	E
■	A	D	R	I	A	N	■	T	A	I	L	O	R	■
F	A	N	■	S	P	I	N	D	O	C	T	O	R	S
B	R	O	■	T	U	T	■	A	N	N	A	T	T	O
I	T	S	■	S	P	Y	■	B	E	E	■	S	O	N

PUZZLE SOLUTION 28

F	T	O	P	■	A	L	A	N	D	■	G	P	A	S
A	R	N	O	■	S	O	P	O	R	■	R	A	L	E
C	A	E	N	■	O	C	T	A	L	■	A	R	A	T
E	L	I	Z	A	B	E	T	H	A	R	D	E	N	■
R	E	D	I	G	■	■	O	S	O	S	■	N	P	R
S	E	A	■	A	S	U	■	■	T	A	T	A	S	■
■	■	A	R	C	T	I	C	C	■	R	A	G	A	■
■	P	O	I	S	O	N	E	D	A	P	P	L	E	■
D	O	U	R	■	W	E	L	L	Y	E	S	■	■	■
R	U	T	H	S	■	■	I	S	E	■	A	R	I	■
S	N	L	■	U	S	F	L	■	L	T	G	E	N	■
■	D	I	R	E	C	T	O	R	S	S	H	O	U	T
S	A	N	O	■	A	L	M	A	Y	■	O	R	N	E
A	G	E	R	■	R	E	A	R	S	■	R	A	E	R
C	E	D	Y	■	S	E	N	A	T	■	S	S	S	S

PUZZLE SOLUTION 29

```
L U V S . B O L A . W S T O N
O N I T . A L A D . I O W A S
A R R R . Z A C H . L L A M A
F I T A S A F I D D L E . . .
. G U S T A V . R A M R O D
. . S L R . E R O . N A P A
L E K E U . O N E P M . N E N
S H A R K S R I V A L G A N G
A L L . E H U D S . C O C A S
T U E S . A B S . L A D . .
S A S S E D . C O R S E T .
. K E E P S F O R P L A Y
S Y B I L . F U J I . E S N E
M E I N E . C S N E . E A T A
L E G S D . S A L S . D S O S
```

PUZZLE SOLUTION 30

```
F D I C . H O S P . K A R E N
R I P A . O P E C . A P A G E
E A R E D S E A L . N I N E R
O N A N E . N O A D S . D S T
N E Y . J A C O B J A V I T S
. . C A D I Z . S N I .
T I T I V A T E . . V A T S
M E R C U R Y . C H A I R E D
S L U E . D A Y G A M E S
. . R B S . O R M A N .
I M P O S I T I O N S . G P A
C A L . T T T T T . S O L A N
T R A L A . O N E L I N E R S
U N S E R . P O N E . C E C E
S E M I S . S W E D . E S E L
```

PUZZLE SOLUTION 31

```
L U P S U . R P I . R A H S
S T U P S . E U L A . E T A H
D I N A H . C M L I . A E R I
. G R E A T P Y R A M I D S
. K R A . W R E T C H
W I L L I A M B O A R D .
H E M E N . E L E V S . L A M
Y O N D . P R O S E . T E T R
S H O . D I C O T . T A C O S
. S E M I D E T A C H E D
U P K E E P . O B I
D A V I D L E E F R O T H
D O A N . E W E R . R U E D E
E L S E . S A R I . E R R O R
R O S S . N Y T . T N A I L
```

PUZZLE SOLUTION 32

```
E L S . A E C I A . S T R E P
A A M . S P I T Z . P O E S Y
S U A . T O A S T M A S T E R
T R L . A N O A . E R E S S O
M A L O R Y S . C R E E .
. H T T M . C H E M . B S S
E N O T E . P O L L E N A T E
B L U E . O H B O Y . A L O P
B A R R Y M O R E . U R A W A
S T S . E A T A . E N C L .
. A W H O . S E T S A I L
C U E S T A . D A Y O . I N I
I N C A R N A T I O N . K B S
N I C H E . D E E R E . A U T
C E L I E . O N R E D . S D S
```

PUZZLE SOLUTION 33

D	O	G	Y			A	G	E	S				R	I	T	E
I	C	A	O			P	L	E	A			T	U	N	E	R
S	C	R	U	M	H	A	L	F				A	P	T	E	R
B	U	G			A	I	D			A	S	L	E	E	P	
U	P	L	A	N	D			B	R	O	K	E	R	E	D	
D	Y	E	D			P	R	I	M					N	E	D
			M	E	L	E	E			B	E	H	E	S	T	
	C	A	M	A	R	A	D	E	R	I	E					
W	A	L	N	U	T			K	O	R	A	N				
A	L	A			E	M	I	T				D	I	S	H	
G	A	M	E	P	L	A	N			D	A	I	N	T	Y	
	B	O	L	E	Y	N			L	A	G			H	A	M
C	A	R	V	E			C	H	O	C	O	L	A	T	E	
A	M	E	E	R			H	I	G	H			E	L	A	N
T	A	D	S			U	V	E	A			E	E	L	S	

PUZZLE SOLUTION 34

O	B	I			W	R	A	P			C	E	S	S			S	I	R
D	E	N			A	U	R	A			A	N	Y	P	L	A	C	E	
E	N	T	W	I	N	E	S			P	A	R	R	O	T	E	D		
S	T	E	R	N			A	S	S	O	C	I	A	T	I	O	N		
			R	E	S			L	E	K			T	A	W			A	V	E
A	L	P	A	C	A			I	T	S			L	U	T	E	S		
W	A	L	K	O	F	L	I	F	E			R	E	V	E	R	S		
E	R	A			T	E	E	O	F	F				A	R	E			
S	K	Y	E			A	I	D			L	I	P			A	C	M	E
			P	E	R			I	R	O	N	I	C			L	A	Y	
R	E	M	I	N	D			D	I	N	N	E	R	T	I	M	E		
E	L	E	C	T			R	E	P			R	E	R	E	A	D		
F	O	G			R	O	E			E	O	N			V	A	N		
O	P	A	L	E	S	C	E	N	C	E			A	C	T	O	R		
R	E	H	E	A	T	E	D			T	I	P	S	T	E	R	S		
G	R	I	T	T	I	N	G			E	G	I	S			L	E	V	
E	S	T			S	A	T	E			T	H	E	E			E	S	P

PUZZLE SOLUTION 35

P	R	O	P			A	G	G	R	O			H	E	L	M
H	E	R	E			L	O	R	A	L			A	S	O	K
I	N	R	E	A	L	T	I	M	E			N	E	S	T	
S	E	S			S	I	D	E	B			S	D	L	T	G
			N	S	A			G	O	L	D	W				
F	R	A	I	D	S	O			S	A	N	R	E	M	O	
L	A	R	C			K	N	X			P	A	I	S	A	N
A	R	E	E	D			T	L	C			H	T	T	P	S
R	I	S	Q	U	E			S	U	B			T	A	L	E
E	N	O	U	N	C	E			T	Y	P	E	S	E	T	
			A	S	I	C	S			H	E	N				
S	P	I	L	T			Z	A	L	E	S			M	I	O
M	U	T	I			G	E	R	I	A	T	R	I	C	S	
I	G	H	T			O	M	A	R	R			O	T	O	E
T	H	E	Y			E	A	T	A	T			G	E	N	T

PUZZLE SOLUTION 36

A	P	A	T	H			T	A	L	E				C	I	D
C	S	P	A	W			R	R	U	N			W	H	O	A
H	A	P	P	Y	J	U	I	C	E			O	A	T	S	
E	L	I	A			E	E	L	Y			S	R	I	A	H
S	M	A	S	H	E	S			O	A	M					
			U	P	T	O	W	N	G	H	O	U	L			
H	O	O	H	A			D	B	C	O	O	P	E	R		
O	S	S	E			P	L	O	C	E			L	A	L	O
T	H	A	N	K	Y	O	U				H	E	L	E	N	
W	A	S	H	E	R	D	R	Y	E	R						
			O	P	E				O	N	E	T	I	M	E	
E	N	N	U	I			B	A	U	D			O	L	E	A
G	T	O	S			T	A	T	T	O	O	I	S	T	S	
G	O	R	E			P	I	A	O				P	L	A	I	T
O	F	A			I	N	D	O			S	S	S	S	S	

PUZZLE SOLUTION 37

I	L	L	■	S	A	R	O	D	■	H	D	A	Y
Z	I	T	I	■	A	R	A	C	E	■	O	R	N	O
O	N	Y	M	■	Y	E	N	T	A	■	M	E	N	D
D	A	R	N	G	O	O	D	■	L	I	E	G	E	S
■	■	O	N	K	■	B	R	E	W	S	■	■
S	O	F	T	C	■	S	A	E	■	O	R	B	I	T
C	L	A	R	■	S	E	T	H	S	■	I	O	N	A
R	E	T	E	■	O	A	T	E	S	■	G	I	G	I
I	O	W	A	■	B	E	H	A	N	■	H	T	E	N
P	S	A	L	M	■	A	E	R	■	N	T	E	S	T
■	L	B	A	R	S	■	F	B	I	■
M	A	G	Y	A	R	■	A	T	L	A	N	T	I	S
P	I	T	H	■	E	E	N	I	E	■	O	H	M	E
A	R	E	E	■	U	N	D	E	R	■	N	O	A	A
A	P	A	R	■	P	E	L	T	S	■	M	T	T

PUZZLE SOLUTION 38

H	A	J	I	■	I	O	N	S	■	E	C	A	R	D
A	N	U	N	■	C	F	O	S	■	C	A	T	E	R
I	N	L	I	N	E	F	O	R	■	O	R	A	M	A
L	I	I	■	O	T	A	■	S	A	N	D	B	O	X
S	V	E	L	T	E	■	T	O	I	■
■	A	T	A	C	O	S	T	■	O	P	A	L
E	T	H	N	O	■	I	S	A	A	C	■	U	R	O
B	L	U	E	B	E	R	R	Y	M	U	F	F	I	N
O	M	S	■	E	R	R	I	S	■	E	X	F	B	I
L	A	K	E	■	N	O	C	A	R	B	S	■
■	C	C	S	■	H	A	P	P	E	R
S	I	L	E	N	T	W	■	T	E	L	■	I	V	O
I	A	L	S	O	■	I	N	S	U	L	A	T	O	R
A	M	B	I	T	■	S	C	U	M	■	T	O	K	E
M	A	S	S	E	■	C	O	P	Y	■	M	N	E	M

PUZZLE SOLUTION 39

E	L	I	S	A	■	T	H	U	S	■	F	E	M	S
B	E	N	I	N	■	Y	E	S	T	■	R	A	Y	E
B	R	I	T	T	■	C	H	A	I	■	O	N	D	E
■	T	H	E	C	H	E	R	R	Y	M	O	O	N
A	L	I	■	N	C	O	■	I	L	L	G	O
G	I	A	N	N	I	■	M	A	U	N	A	■
G	A	L	E	A	■	Z	I	G	S	■	L	E	U
I	N	E	V	E	R	E	A	T	S	N	A	I	L	S
E	A	D	■	A	B	M	S	■	E	S	K	E	R
■	S	E	R	U	M	■	B	A	D	E	N	D
S	H	E	E	P	■	N	A	T	■	A	A	A
C	A	T	C	H	E	S	A	B	R	E	A	K
A	R	T	R	■	D	O	D	O	■	N	S	I	D	E
B	R	A	E	■	A	N	I	M	■	E	T	N	A	S
S	Y	S	T	■	M	Y	O	B	■	D	I	G	I	T

PUZZLE SOLUTION 40

P	R	O	T	E	M	■	I	S	E	E	■	P	C	B
L	I	P	A	S	E	■	S	V	E	N	■	U	E	A
A	V	E	R	S	E	■	U	E	N	D	■	F	R	S
■	M	O	R	A	L	L	Y	S	A	F	E	R
R	O	C	A	■	D	I	T	■	E	S	S	A
I	N	O	C	U	L	A	■	E	S	P	O
N	E	O	■	R	O	N	I	■	C	I	N	E	M	A
T	A	K	E	I	T	O	N	T	H	E	S	H	I	N
T	R	E	X	E	S	■	S	A	M	L	■	U	K	E
■	U	L	A	N	■	P	O	S	A	D	A	S
M	O	H	R	■	A	P	O	■	I	S	N	T
R	U	B	B	E	R	P	E	N	C	I	L	■
B	T	U	■	L	E	E	R	■	H	A	I	L	T	O
I	R	R	■	B	E	R	M	■	U	S	N	E	W	S
G	E	N	■	E	L	Y	S	■	B	I	G	A	P	E

PUZZLE SOLUTION 41

I	S	B	U	T		C	N	B	C		D	O	A	N
O	T	E	R	I		H	E	L	I		I	N	S	O
O	R	A	N	T		E	R	I	T		D	E	L	L
F	A	M	I	L	Y	N	A	M	E		A	M	O	O
			N	E	E		P	E	R	H	A	P	S	
B	E	L	G		S	A	C	S		I	O	N	E	S
B	L	O		U	M	B	O		B	O	R			
S	A	L	E	S	A	S	S	I	S	T	A	N	T	S
		R	O	N		M	R	I	S		T	O	S	
P	R	O	O	F		G	O	A	D		P	H	O	N
A	I	R	T	A	X	I		E	L	E				
R	A	N	I		A	D	D	I	S	A	B	A	B	A
S	T	A	C		N	E	O	N		I	B	E	A	M
E	A	T	A		D	O	T	O		C	L	O	U	D
E	S	E	L		Y	N	E	Z		S	E	N	D	A

PUZZLE SOLUTION 42

W	I	C	K		G	L	I	B		O	R	I	E	L
A	T	L	I		A	O	N	E		R	A	S	P	Y
D	T	E	N		Y	O	D	A		T	B	A	L	L
S	O	A	K	E	D	T	O	T	H	E	B	O	N	E
		Y	M	A			T	G	I					
F	T	S		A	R	T	I	S	T	A		M	O	I
T	H	I	N	G		O	N	K	P		P	A	N	N
D	A	N	I	S	H	P	H	Y	S	I	C	I	S	T
I	N	E	S		E	E	E	E		S	P	D	E	R
X	X	X		M	A	R	R	Y	M	E		S	T	A
		T	A	R			O	E	O					
S	E	N	S	I	T	I	V	E	N	A	T	U	R	E
T	R	I	A	D		N	E	R	I		O	N	I	N
A	A	N	D	E		S	I	M	S		W	D	E	D
S	T	A	I	N		P	L	A	T		N	E	N	E

PUZZLE SOLUTION 43

N	B	A	E	R		S	B	A	C	K		C	C	S
A	A	N	D	E		P	I	T	H	Y		I	A	T
T	R	A	D	I	T	I	O	N	A	L		T	R	U
	R	A	N	A	T		N	E	A	R	E	R		
L	A	C		E	D	E	N	S		M	O	L	D	
O	R	H	E		A	D	U	L	T	S	O	N	L	Y
S	L	I	M	E		S	O	D	O	M				
T	O	C	E	L	I	A		B	S	N	E	S	T	S
	R	E	B	U	S		O	N	M	A	N			
U	S	C	I	V	I	L	W	A	R		T	O	T	E
N	U	I	T		D	A	R	C	Y		O	S	E	
P	E	T	I	T	F		A	V	E	R	T			
L	A	I		M	A	R	X	B	R	O	T	H	E	R
U	N	E		A	I	O	L	I		W	E	E	N	Y
G	E	S		C	R	E	S	C		S	E	D	G	E

PUZZLE SOLUTION 44

M	I	N	T		A	M	E	N		A	E	R	I	E
E	N	C	E		S	A	M	I		N	L	E	R	S
O	C	A	N		S	U	P	S		O	L	D	E	N
W	R	A	P	P	A	R	T	I	E	S		O	S	E
		E	S	S	A	Y		G	M	A	C			
M	Y	R	R	H	S		S	H	I	F	T	E	D	
A	A	E		C	I	S	C	O		A	L	O	N	E
S	W	A	B		N	A	F	T	A		E	B	O	L
H	E	S	S	E		L	L	O	S	A		E	L	F
A	D	S	A	L	E	S		T	H	I	R	S	T	
	E	W	E	R		O	G	E	E	S				
A	E	S		C	E	L	L	A	R	D	O	O	R	S
H	A	S	I	T		C	D	L	I		M	T	E	L
A	T	E	A	R		T	E	A	C		E	T	L	A
B	A	S	S	O		O	R	S	K		R	O	O	T

PUZZLE SOLUTION 45

D	A	V	A	O	■	R	E	S	E	■	L	T	D	S
I	M	O	F	F	■	E	R	T	E	■	A	R	E	E
P	O	L	I	T	I	C	I	A	N	S	G	O	A	L
S	I	T	E	■	H	O	E	R	■	B	G	I	R	L
■	■	L	E	N	I	■	V	A	C	A	■	■	■	■
S	E	A	D	W	E	L	L	E	R	■	R	P	M	■
O	M	N	■	E	N	E	S	■	T	O	D	A	Y	S
S	A	T	Y	R	■	D	A	U	■	T	S	A	R	S
O	G	R	E	S	S	■	T	R	E	O	■	V	O	S
■	S	A	S	■	O	B	S	E	S	S	I	O	N	S
■	■	S	O	L	E	■	T	P	E	R	■	■	■	■
O	S	L	I	N	■	A	S	H	Y	■	E	N	D	O
Q	U	A	R	T	E	R	P	A	S	T	F	I	V	E
P	R	I	E	■	M	O	A	N	■	C	U	R	D	Y
H	E	R	E	■	I	N	S	E	■	I	L	O	S	E

PUZZLE SOLUTION 46

S	I	C	S	■	S	M	E	E	■	O	S	K	A	R
O	T	R	O	■	T	O	A	N	■	V	A	L	L	O
U	S	E	D	C	A	R	S	A	L	E	S	M	E	N
L	A	M	A	R	R	■	Y	M	I	R	■	N	E	I
S	T	A	■	I	M	O	■	I	L	K	A	■	■	■
■	■	■	L	B	A	R	S	■	T	I	R	A	N	A
U	S	M	A	■	P	A	C	K	■	L	T	G	O	V
S	H	A	R	K	S	R	I	V	A	L	G	A	N	G
F	E	R	I	A	■	E	F	A	X	■	U	R	E	S
L	A	M	A	Z	E	■	I	S	H	A	M	■	■	■
■	T	O	V	A	■	S	A	M	■	T	O	I	■	■
R	G	S	■	O	A	T	Y	■	N	O	R	E	S	T
P	E	T	E	T	H	E	M	E	D	I	O	C	R	E
T	R	I	N	I	■	S	H	E	L	■	S	T	I	R
S	E	T	T	E	■	T	A	R	E	■	S	A	C	S

PUZZLE SOLUTION 47

C	B	S	T	V	■	R	A	R	E	■	C	H	A	D
R	U	M	B	A	■	I	D	E	A	■	A	A	R	E
A	D	R	A	G	■	A	D	D	R	■	V	I	T	A
M	S	F	■	R	O	L	L	I	N	G	I	N	I	T
■	■	P	A	T	T	■	I	M	A	G	E	D	■	■
K	E	V	I	N	C	O	S	T	N	E	R	■	■	■
E	L	I	S	T	■	L	I	G	N	■	D	O	M	■
N	O	T	A	■	F	L	O	S	S	■	S	O	I	T
O	I	O	■	A	L	E	A	■	P	H	I	S	H	■
■	■	P	A	Y	I	N	C	R	E	A	S	E	S	■
T	S	G	A	R	P	■	R	E	P	T	■	■	■	■
W	R	A	P	P	A	R	T	I	E	S	■	S	S	A
E	T	T	E	■	P	E	E	S	■	I	N	M	A	N
R	A	E	R	■	E	A	R	P	■	C	O	A	T	I
E	S	S	Y	■	R	O	M	Y	■	O	R	N	O	N

PUZZLE SOLUTION 48

B	E	A	S	■	E	R	R	■	C	G	T	W	U
O	P	T	O	■	H	E	H	E	■	A	R	H	A	T
S	L	O	B	B	E	R	E	D	■	T	E	R	R	A
■	N	Y	A	L	A	■	O	R	A	N	G	I	S	H
■	■	O	E	R	■	■	M	I	O	■	■	■	■
■	T	H	E	O	R	I	E	N	T	P	R	E	S	S
B	R	O	O	M	■	Y	E	A	H	S	■	B	U	G
L	I	M	N	■	F	A	R	B	E	■	T	O	M	T
A	S	I	■	R	A	D	I	O	■	C	O	O	P	S
T	H	E	M	S	T	H	E	B	R	E	A	K	S
■	■	E	V	S	■	■	S	H	R	■	■	■	■
L	O	V	E	P	O	E	M	■	E	A	N	O	L
I	S	I	T	I	■	D	R	U	M	M	A	J	O	R
R	I	V	E	N	■	G	E	N	E	■	O	X	U	F
R	E	O	R	G	■	E	S	E	■	H	G	T	S

PUZZLE SOLUTION 49

```
A L P H   L E G S D   O R C A
V E R A   O N A N E   L I M N
C M I I   A T R E E   D D A Y
L O O K S F O R A J O B
U N R U H   K A R A O K E
B Y S   I R V S   Y S I D R O
    C R E A M S   L O I N
  F O U R C Y L I N D E R S
L I R R   A E R I F Y
O N L Y I F   D E L I   A D J
K E E P S A T   L A M I A
    L O V E S C O M P A N Y
G O T A   O P E N H   S L A V
N O R N   R I P E N   E I R E
U P A T   S D A T E   S E S E
```

PUZZLE SOLUTION 50

```
G H E E   W O R D S   M C L
M A R X   I S E R E   I E R E
A I D E   L T C O L   T R A G
T R A M P L E D   V C H I P S
    P E Y O   F A C E T E D
H E L L N O   A G U E
G R E A T U N C L E   L A T
T E T R A   E S L   B B A R T
V I S   E L A S T I C I T Y
    G E N L   U N O C A L
A C Q U I R E   M N E M
T H A T S O   T E N T P E G S
B I N E   B A S I E   L I M O
A D D N   E X C E L   E N A S
T E A   D E A R S   X E N O
```

PUZZLE SOLUTION 51

```
A L F I E   P I U S   D L I
T O I L S   T A S K S   E O N
W O R L D S E R I E S   C R A
O K E Y   O A S T   B O N N
  G R O W M I S T Y E Y E D
F U R I O S O   A T O E
N C A A S   P F E N N I G S
M A C   B A S E N   N O I
A L E A T O R Y   A T T A M
  N O R M   S T R E E T S
Y A N K E E C L I P P E R
E N Y A   H U N K   S F P D
L I A   C H A R G E A H E A D
L O C   A M I E S   D O R I A
O N K   T O R R   M T E L Y
```

PUZZLE SOLUTION 52

```
A B B A S   G I F S   A N A T
M O O R E   O N L Y   N O L O
B A Y E R E S S E N T I A L S
I T O O   T H E X   O M A R S
    L I T E   I A M A
B R E A D A N D B U T T E R
S E T S A   E L S   O L E S
A R Y   H A M L E T S   I S A
W O M B   W A C   A C H E S
  W A T C H Y O U R M O U T H
    W L O O   N E L L
H E R E I   R E D D   L S T S
S H E L O V E S Y O U Y E A H
O L A V   I S M E   B E T H E
N O P E   M S E D   U R I E L
```

PUZZLE SOLUTION 53

R	I	D	E		O	T	H	E		D	U	S	T	S
U	N	E	S		P	E	O	N		A	L	L	O	K
M	A	N	I	F	E	S	T	D	E	S	T	I	N	Y
I	N	S	T	I	R		L	A	S	H		C	E	E
		X	A	S		S	S	I		E	R	Y		
N	O	G	R	E	T	A	S	H	A	K	E	S		
I	N	R	E		O	L	E		I	B	I	S		
S	K	A	T		R	E	E	V	E		O	N	A	T
	P	F	C	S		T	E	D		L	T	D	S	
	T	H	E	D	A	V	I	N	C	I	O	D	E	
P	E	U		Q	O	M		N	A	W				
I	A	N		U	S	A	F		B	T	E	N	E	D
P	R	I	C	E	O	N	O	N	E	S	H	E	A	D
I	N	O	I	L		D	U	H	S		U	R	S	A
T	A	N	I	A		A	R	A	T		D	O	T	Y

PUZZLE SOLUTION 54

U	N	C	A	P		Z	E	I	T		H	A	V	A
S	A	C	R	A		I	L	S	A		I	S	A	R
M	M	I	I	I		N	A	S	I		D	E	L	T
	E	X	C	U	S	E	M	E	P	L	E	A	S	E
		H	T	O			L	E	I	A				
A	A	S		E	N	O	L		I	T	W	A	S	I
S	W	O	B		O	H	E	D		E	A	S	T	S
S	A	T	I	S	F	I	E	D	O	R	Y	O	U	R
N	I	O	B	E		O	D	A	S		S	U	P	E
S	T	O	L	G	A		S	Y	M	S		R	E	D
		E	N	D	E			A	O	W				
C	O	M	B	I	N	E	D	I	N	C	O	M	E	
O	F	M	E		E	L	Y	S		K	O	I	N	G
N	U	L	L		X	E	N	A		E	D	D	I	E
E	S	I	T		A	R	E	W		T	S	I	D	E

PUZZLE SOLUTION 55

P	O	P	E	S		A	B	I	E		N	P	I	N
U	C	L	A	N		T	R	O	N		I	N	T	O
N	E	A	T	A	R	R	A	N	G	E	M	E	N	T
J	A	N	A		E	A	G	E	R	E	R			
A	N	E		B	U	L	G		K	O	M	B	U	
B	A	T	A	A	N		A	L	A	D	D	I	N	
		P	R	I	N	G	L	E		X	T	C		
	U	U	U	I	T	O	R	L	O	S	E	I	T	
P	O	S		E	T	E	R	N	A	L				
O	F	F	B	A	S	E		A	U	G	U	R	S	
R	A	L	L	Y		E	A	R	L		R	A	T	
	O	E	R	S	T	E	D		K	A	T	O		
S	E	C	U	R	I	T	Y	C	O	U	N	C	I	L
A	S	A	S		P	A	P	I		L	A	I	N	G
G	A	P	E		A	T	E	A		A	P	L	E	A

PUZZLE SOLUTION 56

A	P	R	S		O	G	O	D		D	W	E	E	B
S	H	U	E		F	O	R	E		O	H	A	R	A
M	I	M	E		G	E	R	M	I	N	A	T	E	S
A	L	O	N	S	O		O	D	E		A	I	S	
D	I	R	T	Y	D	A	N	C	I	N	G			
	P	R	O	M		V	E	R		N	I	G	H	
T	R	U		O	N	O	R		I	A	M	N	O	W
W	O	M	E	N	S	W	E	A	R	D	A	I	L	Y
A	T	O	S	S	A		I	C	E	D		T	D	S
S	H	R	E		E	D	D		A	S	I	E		
	L	O	V	E	S	C	O	M	P	A	N	Y		
R	A	J		U	A	L		I	S	I	T	M	E	
A	L	E	R	T	R	E	B	E	L		T	I	E	A
K	A	F	K	A		R	E	A	O		E	V	A	N
I	N	F	C	T		S	E	L	F		S	E	N	S

PUZZLE SOLUTION 57

PUZZLE SOLUTION 58

PUZZLE SOLUTION 59

PUZZLE SOLUTION 60

Puzzle Solution 57
```
Y A H W E H   M S R P   R T S
O N I O N Y   O W E S   E A T
D E C O D E   A A C Y   R K O
    D O N A T B E C R U E L
V J J D W A G     H I N T A
U E A   S A E   R I D S O F
E F T S     T P I E C E
  F O U L G H O U L S O U L
    D O R A D O     N S E C
S A Y S S O   E E K   P T A
A P O E T     P A R T S A N
N E W S A G E N C I E S
C M L   R O E G   S I E G E S
H A E   T O R O   E N T E R O
O N D   S N O R   R A S T A S
```

Puzzle Solution 58
```
E S S U   S W A P     M E H
S H A N T   H I T O R   A S O
T A K E S T O T A S K   D S T
E R S A T Z   T R E E N A I L
    S O A S     A L O G
U T T E R R O T   S L E A Z E
S U E   M S T A R   Y A S I R
E M A N   O N E   R C P T
T O P I C   L I A R S   A P E
O R A C H E   A L L E G R O S
    R H I N   M E G A
P E T E R S O N   S U M M I T
T G I   R E V E R S E B I D S
A A E   S A U R Y   D O R I C
H N S   T M A N   L A D A
```

Puzzle Solution 59
```
A P E A R   R I A   S F P D
R E N T E   C O M B   A A R E
D A V I D B O W I E   M L I N
  C H I C A N E   A S C I
U S C   O D O N   P R E Y S
S H A I T A N   A L A R
R E T O P   U H U R A   E O E
D I S C O N T E N T I S T H E
A K C   K A S E M   N O U S E
  A C E H   A L S O R A N
W I N E R   A P I P   N Y S
M O N A   O L D P R O S
E N E S   S A U E R K R A U T
A I R E   E K E D   E A R N S
T A S S   T E L   N S I D E
```

Puzzle Solution 60
```
S A M O A   R T E S   L A D E
B C E L L   E R N A   O N O R
C D G E L   G O O F   V T E N
  C R O C O D I L E T E A R S
    L A T   T A M E S T
C R O S S T H E L I N E
S O R E H   S W E E K   D O W
U D O N   T I E R S   M O N O
P E S   M E N L O   A E T N A
  D A N G L I N G S H A D
A P R I Z E     E H S
F U L L Y R E C O V E R E D
A R E A   I T A T   E O S I N
L E S T   F C T O   L O T T A
L E S E   E S O S   S M A S H
```

PUZZLE SOLUTION 61

```
C A S U A L S . . N E H R U
A N O S M I A . D R I V E I N
R O U T I N G . R E S E A L S
A T T I R E S . E L A S T I C
T H A N . . . S A N . H E R .
S E N O R . R U S T . E V E .
. R E V E R E N C E S . N O W
. . D E P L O R E . . . . .
A D S . S O L I D S T A T E .
R A T . R I T E . S T R A Y .
R Y E . S I C . . S E R E .
I C E D T E A . S M E T A N A
V A R I A N T . A E R A T E S
E R E M I T E . R A N K E S T
R E D E D . . I N S E R T S .
```

PUZZLE SOLUTION 62

```
F T H S . . L O N G B O A T
R A U L . P A N A T E L L A
O G E E . E A V E S D R O P S
M O V E S A H E A D . C F O S
M N O P Q R . N L A K E . .
. S Y N O D . Q L U P F H
S I S . D E N E B . U S A I D
L E A H . R E R O W . E S S A
Y O G A S . I S A A K . T H Y
S H O R T U . A C C E L . .
. D U L A C . K N A V E S
I L I A . S C H O O L Y A R D
F A N L E T T E R S . M L I N
I N T E R E S T S . E L E A
M A D E I R A S . N O S H
```

PUZZLE SOLUTION 63

```
W E L C O M E M A T . I V O R
I N A C T I V A T E . W E R E
C O N S O N A N T S . A S I S
C U D . H I S N . I S P E P
A G R A . . A B A M . U N O
. H U S H U P . A T P . C T N
. S E L E C T O R . C A S
C L I E N T S C A S E F I L E
H E N . C R E V A S S E .
E C H . O A T . N A S D A Q
E T A . O S A Y . . S E U L
T U L I P . A A R P . N A A
A R A L . H E R B A L P E R T
H E N S . A N N I H I L A T E
S R T A . P O S T S E A S O N
```

PUZZLE SOLUTION 64

```
W E L C O M E M A T . I V O R
I N A C T I V A T E . W E R E
C O N S O N A N T S . A S I S
C U D . H I S N . I S P E P
A G R A . . A B A M . U N O
. H U S H U P . A T P . C T N
. S E L E C T O R . C A S
C L I E N T S C A S E F I L E
H E N . C R E V A S S E .
E C H . O A T . N A S D A Q
E T A . O S A Y . . S E U L
T U L I P . A A R P . N A A
A R A L . H E R B A L P E R T
H E N S . A N N I H I L A T E
S R T A . P O S T S E A S O N
```

PUZZLE SOLUTION 65

```
S E E M T O B E ▮ ▮ ▮ I S L A
T R A L A L A L A ▮ K S T O N
U R T I C A R I A ▮ A R E M Y
D A M N ▮ R E A I R ▮ E A L
S S E ▮ F B I ▮ A M A R L L O
▮ E P I S C ▮ E T A M I N
D E C A M E T E R ▮ S T I N G
R A O M ▮ S E R A C ▮ O L D E
A R O O M ▮ R E T I C U L A R
F T K N O X ▮ S P R I T ▮
T H I N N E R ▮ A E C ▮ W B A
S T E ▮ I S E N T ▮ J I L L
M O J O S ▮ P O R T F O L I O
A N A C T ▮ O B O E R E E D S
N E R D ▮ S L E E L E S S
```

PUZZLE SOLUTION 66

```
I N G E S T I O N ▮ B O L A
S A N I T A R I A ▮ A B A C O
S T O R Y L I N E ▮ R E R A N
H U M E ▮ E S K ▮ S O I G N E
O R E ▮ E N E S C O ▮ S E T H
T E S ▮ I T S ▮ H I B A C H I
▮ A L S ▮ A E R O N A U T
C E R E ▮ R T E ▮ P C P S
S H A R E D I N ▮ P E E ▮
T O R O N T O ▮ A R E ▮ M M M
A R A W ▮ E T H I O P ▮ O O O
T U C K I N ▮ A L B ▮ I N S P
U S H E R ▮ T R I A L D A T E
S E E Y A ▮ S E N T E N C E D
▮ S S S S ▮ E D G E T O O L S
```

PUZZLE SOLUTION 67

```
B E B E ▮ T R A G ▮ I N D U S
I L A Y ▮ I O N E ▮ N O E N D
T H R E E M E N I N A T U B A
T I M B R E ▮ I S A N ▮ C I E
▮ E T O ▮ E R I ▮ E D S
F A S H I O N P L A T E S ▮
R I P A ▮ G E E ▮ Y A W P
O W E N ▮ O S R I C ▮ P I E S
A C D C ▮ O R O ▮ O L A N
T H A N K N O W L E D G E
S B A ▮ B Y O ▮ C B Y ▮
L A C ▮ A P R I ▮ E R A S E S
I S L A N D E R S L A N D E R
C I E R A ▮ A M U L ▮ D A R A
E L S A S ▮ N A B S ▮ S K Y S
```

PUZZLE SOLUTION 68

```
B I R S E ▮ P S A T ▮ U S C G
A T P A R ▮ P L U A ▮ N O H O
B O M B A R D A D R A B M O B
▮ R U E ▮ I I I I ▮ M I S
▮ F R E Q U E N T F L I E R
M E E ▮ S N L ▮ S A I S ▮
A R F S ▮ E O N ▮ N I N O S
P A R K E D I L L E G A L L Y
S L Y A S ▮ Q I X ▮ H E D S
▮ L T D S ▮ M I L ▮ R I T
U D D E R N O N S E N S E ▮
O N A ▮ E E O C ▮ T S E
R A T E S A R T E S T E A R S
A R E D ▮ D E E P ▮ E L I A H
L Y R A ▮ S R T A ▮ R Y D E R
```

PUZZLE SOLUTION 69

```
D E W S . E R I C . H I G H S
A G A I . M E T H . A C T S A
H O R T . E P H A . S A D T O
L I M A O R L E N T I L . . .
S S T . A G A . D E D . H I S
. M O N K E Y B U S I N E S S
. . B Y S E A . S C O W L S .
A T I C . . D N C . E N E S .
B I S T R O . J E O F N . . .
C A L V I N C O O L I D G E .
S S S . P A Y . P D S . A N S
. . W I N N I E T H E P E W .
A V I A N . I N R I . L I S I
V I N G T . C R A M . E N C S
O R F E O . S E S E . E G O S
```

PUZZLE SOLUTION 70

```
E Y E B . S H R E . L U C A S
C E L O . N I E R . E R A T O
H A S A C A T C H . E S P O O
T H E T O P . S A A R I N E N
. . . E L S A . R L S . . . .
. C O L D H A R D C A S H E W
R E M . S O R A . . T R A L A
I R A S . T P O D A . O J O S
A A H E D . C E B U . J A I .
A M A Z I N G K R E S K I N .
. . . L B O . N T H S . . . .
S L A M B A N G . T E T R A D
A U G I E . U N A I R A B L E
K N O R R . T M A N . R I F F
I N G O T . S A R G . S S S S
```

PUZZLE SOLUTION 71

```
A S L . I R A N T . E D A T E
S H E . N O S E R . N E V I N
T E N . T Y C H O . B W A R D
O L D F A S H I O N E D . . .
P F L U G . . P Y R R H I C .
. . R L E S S . E G O I S T .
M T T . I L L I N . P F U I .
A I R T O A I R M I S S I L E
U T A H . T E E N Y . S I S .
R E C E S S . E X A M S . . .
A R I E T T A . . M E R I T .
. . N E E S O N N E P H E W .
M E N D A . A T E A T . E R I
A R O O M . H O L B R . T E L
S E R F S . I S L E Y . T I L
```

PUZZLE SOLUTION 72

```
S M E W . A S A B C . L I T E
H A S H . T O S A Y . A D O S
A S T I . R O W E R . N I M S
. C O M B I N E D I N C O M E
. . W R A Y . E L I E . . . .
D A C H A . I L K . K D C F E
E N C A G E . O E N . L I A .
I D E M A N D A R E C O U N T
S M L . C E T . C O H E I R .
M Y L A R . B H A . N I S S E
. . N O A A . T S G T . . . .
S T A N D S T O R E A S O N .
E R M A . P I L O T . Y I P S
P E E L . I N D I A . O L I N
T O R S . E G E S T . U S N A
```

PUZZLE SOLUTION 73

```
G O D W E ■ B A S E ■ B A R A
I C A H N ■ A D H D ■ B R A C
T H R E E M I L E I S L A N D
■ ■ N R A ■ E A T A ■ C D C ■
■ T R I G G E R F I N G E R ■
A H A ■ Y I N ■ S N E R ■ ■ ■
N E J D ■ C N N ■ S O O T Y ■
S T I C K S A N D S T O N E S
E A V E D ■ W E A ■ M E N E ■
■ L L D S ■ D N G ■ N O R ■ ■
■ P O L A R I C E C A P O N ■
E A R ■ N I T A ■ T Y E ■ ■ ■
B R I D G E F I N A N C I N G
R E O S ■ S O N O ■ O O N A S
O R N O ■ T R E E ■ R S V P S
```

PUZZLE SOLUTION 74

```
M X L V ■ I C O N ■ E T A G E
I V I E ■ T A T A ■ V I B E S
S I G N ■ U F O S ■ E T A T S
C I N D E R F E L L A ■ C I I
■ ■ E R B E ■ ■ E R M I N E ■
Y E M E N I ■ F E N D I ■ ■ ■
U B S ■ ■ T R A D E L A W S ■
R O G E R O V E R A N D O U T
I N T H E O P E N ■ ■ N S A ■
■ ■ L U N G S ■ A E N E A S ■
C O R O N A ■ S T A D ■ ■ ■
E N E ■ I S O F T E N H A R D
S K I R T ■ A A A A ■ J A T O
T E N S E ■ S K I M ■ E N E R
A Y E A R ■ T E D S ■ P I S A
```

PUZZLE SOLUTION 75

```
M E A D ■ A F T A ■ I D S A Y
I S T O ■ D I A G ■ O E I R O
S C O W ■ A R T E ■ W D G R N
S U N D A Y S U N D A E ■ ■ ■
U D O ■ C S T ■ C I N E A S T
S O F T C ■ R H Y S ■ S Q S ■
■ ■ B E R L I ■ U A T T Y K ■
■ P A S S E D M U S T A R D ■
H A R P S T ■ O N E R S ■ ■ ■
E R I ■ I A M B ■ I K N E W ■
C A P E A N N ■ R S A ■ A M A
■ ■ B E A G O O D L O S E R ■
D I C O T ■ L U K A ■ H A R M
I M E A N ■ I S E E ■ S L I T
S A N T A ■ A T N S ■ O S L O
```

PUZZLE SOLUTION 76

```
I S E E M ■ O C T O ■ T H O M
N O T M E ■ K M A N ■ H U M E
S P L A T T E R E D ■ A L E S
T E A C H M E ■ R E S T A R T
■ ■ ■ O A F S ■ L L O S A ■
G R A N D C E N T R A L ■ ■ ■
U L N A S ■ O A K Y ■ O M S ■
S C A G ■ O S R I C ■ P H E W
H G T ■ O R S K ■ T H I S I ■
■ ■ I N T H E B E D R O O M ■
D I C T U ■ L E A R ■ ■ ■
A T L A S E S ■ D R A G O N S
K A E L ■ S A L E S W O M E N
A G A I ■ T E A C ■ E R N I E
R O T C ■ A S O K ■ R E I N E
```

PUZZLE SOLUTION 77

```
R S T U . E S T D . . T W O A
M A S T E R R I C E . B E L G
N E G O T I A T E S . S A S E
. S T P A T . S L E W . V E T
. . I T U P . L S E V E N .
V I S A S . O K S . A I S .
S O W N . N R A . T V G A M E
O N I . A S T A I R E . W G T
P O S A D A . B O A . D E E D
. H A J . S A N . A B B R S
. P G A C R C . S A B U .
A R U . T A H R . S A S H A
C A A N . G U I L L O T I N E
L Y R E . U S E F I N E S S E
U S D O . S L O P . R T E E
```

PUZZLE SOLUTION 78

```
O L G A . C O A L . L O S T H
H O L M . O N C E . I D E E S
H O A R . Y E O H . B O N D I
I N D I A N A P A C E R .
. . T P E R . R A R . S E T
. L S A T S . . N T U P L E
L A L . S S O R C A Y R E V E
A T A N . N A P . . I W I N
D O M O A R I G A T O . E S A
D Y N A M O . G L A D H .
S A G . P B A . F E A T .
. A B B E Y L I N C O L N
L E O X I . T A U S . A D I O
G E N O A . A N T E . M A R G
S W A N N . T A E L . P S A S
```

PUZZLE SOLUTION 79

```
C A T . A G O G . S C O W . C O T
I C E . B A B U . A R M A T U R E
G R E E N T E A . R E A S O N E R
S E T T O . A N C I E N T N E S S
. . O U R . H O D . P I E . A T E
U P T I M E . R A Y . F E T E S
S A U S A G E D O G . R U D E S T
A R M . L E G U M E . E L D .
F A S T . S O B . N B C . O R C A
. W I T . L U C I T E . E O N
S T E I N S . I N Y O U R F A C E
P E L T S . A N D . M E A D O W
A R E . P A L . U S A . C G I
R E C T I L I N E A L . T O N N E
E N T H R O N E . S A B O T E U R
S C R E E N E R . S T A R . S K I
T E A . R E D O . Y E N S . S E E
```

PUZZLE SOLUTION 80

```
C E L E B . D U D E . A G U E
A M U S E . O N U S . D A R K
D A M P S Q U I B S . D Y N E
S I B . O U S T . A D O .
. L E F T I E . C Y A N I D E
. R A S P . T A I L S P I N
G A Y S . V E R S E . S O D
H E A T . D E P O T . A I D E
O R R . O R I E L . S L E D
S I D E L I N E . A S I A
T E S T I F Y . D R I F T S
. H O T . B A C K . E W E
E L A N . N O R T H K O R E A
R A N I . E R I E . I N A P T
E P I C . T E E D . M E L T S
```

PUZZLE SOLUTION 81

```
H A L F A █ R A M A █ A D A K
O R I O N █ O D O R █ R I F E
N O T O N █ O M R I █ E L L Y
O N E D I M M S I O N A L █ █
R I S █ K O R █ T S O █ I S U
█ P A L A Z Z O P A N T S █ █
M O C A █ A T O █ █ E A G R E
I H O P █ R E N E E █ R E A R
S A N A A █ █ E A R █ G R P S
D R A W B R I D G E A H █ █ █
O E R █ O O S █ L S E █ C B S
█ T R U S S F E I N G O L D █
T R I O █ T H E E █ E I R E A
Y E S M █ R O L Y █ A N N E E
R O T O █ A T I E █ S O U P S
```

PUZZLE SOLUTION 82

```
A S T A B █ E P P S █ S P F S
B A H I A █ M I L T █ Q U I K
O D E R N E I S S E █ U N D E
U A R █ K E G █ R R A T E D █
█ M A J O R G E N E R A L S █
T W O T O █ E N D U S E █ █ █
F A S T B █ O N M E █ A V I █
R Y E S █ U N C A S █ H B A G
S S T █ U S O C █ B U S C H █
█ O V E R H S █ I T S A T █
M I S T E R N I C E G U Y █
A N I T A S █ A D C █ S C S
R A T A █ F A N G T A S T I C
I L A W █ E M A G █ T H E T A
A L T A █ E B B S █ S U M O S
```

PUZZLE SOLUTION 83

```
M T S I N A I █ T B S █ S S A
P A T M O R E █ A A S █ P P D
S T U F F E D S O L E █ O I O
█ D I E M █ O I L █ L I G N
S Y D N E Y A U S T R A L I A
R T E E █ T T T █ E V E R I
A D D █ I E O H █ B G I R L S
█ T H U M B H O L E █ █
R A S H A D █ E U D E █ G U T
I L I A D █ O N T █ Z U L U
P O S T A G E D U E S T A M P
T H E S █ A N I █ D O I N █
I A N █ A M O N T I L L A D O
D O O █ L M N █ K N E E C A P
E E R █ S A E █ L A S S O E S
```

PUZZLE SOLUTION 84

```
A C A S T █ A A B B A █ F I B
M E T O O █ I N R E S E R V E
S L I P P E R Y A S A N E E L
O T E █ G L Y P H █ G E S T
█ █ S U A █ O M A R █ █
A T T E N █ G R A D U A T E D
S H U L █ P O T █ E R V I N E
B O R D E R L I N E C A S E S
A R I O S O █ N U M █ I N R O
D O N M C L E A N █ A L T O N
█ E E L S █ W C S █ █
I R M A █ U T H E R █ B L T
D O O D O O D O O D O O D O O
D I S C O V E R Y █ B U O N O
O L S █ P A D M A █ S I G I L
```

PUZZLE SOLUTION 85

A	S	P	I	R	E	■	M	C	A	T	■	O	S	S
S	E	E	N	A	S	■	I	H	R	E	■	L	A	P
H	E	Y	N	I	N	E	T	E	E	N	■	E	N	L
O	Y	E	■	L	E	X	■	V	A	L	V	A	T	E
R	O	A	M	■	S	O	M	A	■	B	A	S	I	N
T	U	R	O	W	■	O	L	D	■	S	T	A	D	■
■	C	H	E	E	S	I	E	R	■	E	G	O	■	■
L	O	O	K	I	N	T	H	E	M	I	R	R	O	R
E	N	V	■	G	A	Y	P	R	I	D	E	■	■	■
A	L	E	G	■	M	M	I	■	S	A	L	P	A	■
N	O	R	S	K	■	O	T	T	O	■	M	O	H	M
N	A	H	U	A	T	L	■	C	M	A	■	L	O	E
E	N	A	■	P	R	O	H	I	B	I	T	I	O	N
S	T	N	■	P	A	G	E	■	R	R	A	T	E	D
S	O	D	■	A	N	Y	A	■	E	S	S	A	Y	S

PUZZLE SOLUTION 86

C	A	L	X	■	W	I	T	C	H	■	P	R	E	K
E	X	E	L	■	B	A	H	I	A	■	A	S	T	E
C	O	N	V	E	R	S	E	A	L	L	S	T	A	R
E	N	O	■	M	E	S	O	■	I	S	U	L	I	■
■	■	D	E	A	E	■	D	A	M	U	■	■	■	■
M	A	R	A	U	D	■	I	A	L	E	■	B	O	D
U	R	I	M	■	A	S	S	A	Y	■	I	M	A	■
B	R	O	A	D	W	A	Y	D	I	S	P	L	A	Y
L	I	D	■	R	A	M	O	S	■	U	L	N	A	■
E	D	E	■	O	G	P	U	■	R	A	N	G	I	N
■	■	S	I	E	G	■	Y	O	L	K	■	■	■	■
R	E	P	A	D	■	E	A	S	T	■	A	B	C	■
S	L	A	P	S	T	I	C	K	C	O	M	E	D	Y
V	I	S	A	■	I	S	T	O	O	■	O	R	A	L
P	E	E	N	■	T	R	O	V	E	■	T	O	Y	S

PUZZLE SOLUTION 87

R	A	B	B	I	■	O	O	O	H	■	S	C	A	B
A	C	E	I	N	■	U	P	S	Y	■	C	A	T	O
N	O	A	D	S	■	T	E	M	P	■	O	O	R	T
K	I	N	D	O	F	B	L	U	E	■	T	R	A	S
■	N	S	Y	N	C	■	N	S	E	C	■	■	■	■
■	■	■	G	A	M	E	D	■	F	H	O	L	E	■
S	E	R	V	■	R	A	P	■	I	T	E	M	E	D
O	D	I	E	■	D	M	A	S	S	■	G	A	N	G
L	I	B	R	A	S	■	C	P	S	■	G	R	A	Y
I	N	S	T	R	■	P	T	R	A	P	■	■	■	■
■	■	E	D	I	E	■	I	L	I	A	C	■	■	■
G	A	R	B	■	N	E	W	E	D	I	T	I	O	N
P	U	R	R	■	A	K	E	Y	■	N	I	M	O	Y
A	N	N	A	■	N	E	A	R	■	T	S	E	T	S
S	T	A	E	■	E	R	N	E	■	H	I	R	E	E

PUZZLE SOLUTION 88

A	C	T	E	D	■	U	D	O	S	■	N	A	C	L
P	O	E	T	E	■	D	O	H	A	■	A	L	L	E
H	O	N	O	R	B	O	U	N	D	■	I	S	A	W
I	R	O	N	E	R	■	B	O	L	■	L	O	R	D
S	S	R	S	■	Y	S	L	■	E	M	F	■	■	■
■	■	■	A	T	E	■	R	H	I	N	O	S	■	■
H	T	M	L	■	N	U	K	E	■	O	L	A	N	E
S	O	P	H	I	S	M	■	N	X	S	E	N	S	S
O	R	E	A	D	■	P	N	O	T	■	S	A	E	S
N	A	G	S	A	T	■	O	T	E	■	■	■	■	■
■	■	A	S	E	■	N	E	R	■	E	B	R	O	■
A	R	C	A	■	N	T	H	■	R	O	L	L	O	N
B	U	M	P	■	T	H	E	F	A	R	S	I	D	E
A	N	D	S	■	H	A	R	I	■	A	I	D	A	N
S	E	R	O	■	S	I	O	N	■	Y	E	S	N	O

PUZZLE SOLUTION 89

```
P I B B   T O A D   B R A I N
A N O A   I L L E   O U T D O
N T W T   V E E S   D C T E N
D E T H R O N E D   I K N O W
G R O O V E     E S E S
      E D A S   U S A B L Y
R A N D R   D C U P   C R E E
A L O E   H O A R S   K E N S
H U R L   D B L S   A S L I M
M I N I M A   P A S E
    N A Y S     H O T T E A
A C T E D   T R A I N S T O P
T A B A C   O U R N   K I L O
R R S T A   E E E E   E M I L
I S P E P   D R E D   D E C O
```

PUZZLE SOLUTION 90

```
O R F F   A S S E T   D A I L
B O L E   C A S C O   O F M E
E L E V   C B S T V   Y A H I
S L E E P E R S   A L O T O F
      R I S E S   R A U
K E V I N S     L I C   K G B
O V I S   M D C C I   Y O O
L I G H T S U P T H E R O O M
A L O   P A S T O   A T U B
S S R   O O H   U N P O T S
      A D P   A R N O S
A G O R A E   P U S H O V E R
B A B U   D R A N O   N E N E
Y L E M   R U L E R   G G G G
E L Y S   O B E S E   S A R D
```

PUZZLE SOLUTION 91

```
G A U R     I F I   A V A S T
E N N E   M E R E   P I P P A
D A D A   I D E O L O G I E S
    E C O N   R H E   G A N T
D I R T B I K E   G F O R C E
I S P   S M E   R A R   Y E R
D U A N E   N A I L E D
  P R E S B Y T E R I A N S
    U S U A L S   G O Y A S
O L A   E R N   E G H   A D E
D I R N D L   S N I T C H E D
I M A Y   A K U   S S I N
S P R A Y P A I N T   D Y E R
T E A S E   O T O S   E A R N
S R T A S   S E A   R H Y S
```

PUZZLE SOLUTION 92

```
E A R N   E I L A T   I S H E
Z E I T   G N A S H   S T A G
E R O S   O L I N E   L A V E
K O D   P I A N O P L A Y E R
    I F S S W   E R E S I
G R A H A M   T E A O R
R E B U T   S P U R N   M T A
P A L P   W H O R L   S A R S
S O O   P R A D O   O M N E S
    A R E H A   G S E V E N
  G A M E S     C R A E S
R E C E P T A C L E S   W M P
U S E R   L A O A T   T I E R
E T I C   E B S E N   E L L E
R E T E   S A Y S A   E D T V
```

PUZZLE SOLUTION 93

I	N	R	O	■	M	C	M	I	■	■	R	A	D	H
M	A	A	R	■	S	E	U	L	■	H	A	D	O	N
P	I	A	N	O	S	O	L	O	■	A	J	V	O	S
E	R	A	O	F	■	■	T	I	A	R	A	E	D	■
R	O	T	■	S	P	O	I	L	E	D	B	R	A	T
I	B	A	■	T	A	N	■	O	C	H	■	B	D	S
L	I	S	P	■	N	A	V	■	■	I	A	S	S	E
■	■	M	U	S	T	S	E	E	T	V	■	■	■	■
D	I	D	S	T	■	■	H	U	R	■	E	R	A	T
A	M	E	■	I	N	H	■	L	E	D	■	E	T	S
O	P	H	E	L	I	A	P	A	I	N	■	H	E	H
■	L	O	G	I	C	A	L	■	■	A	C	E	L	A
G	A	R	R	S	■	G	O	T	O	S	L	E	E	P
A	N	N	E	E	■	E	W	E	R	■	E	L	S	E
M	T	S	T	■	N	S	E	C	■	■	A	S	S	D

PUZZLE SOLUTION 94

E	C	C	E	■	E	L	H	I	■	H	S	I	N	G
A	R	O	D	■	X	I	A	N	■	O	N	T	O	E
T	E	R	A	■	U	S	A	C	■	W	A	S	O	N
S	A	D	■	S	L	I	G	H	T	S	K	I	R	T
I	M	I	G	H	T	■	■	R	A	Y	■	■	■	■
■	L	O	A	■	U	S	T	E	D	■	A	R	M	■
O	I	L	O	N	■	P	C	T	S	■	Y	L	E	M
S	H	E	E	T	S	T	O	T	H	E	W	I	N	D
H	A	R	Y	■	H	O	O	T	■	M	C	M	I	I
A	D	A	■	L	E	A	P	T	■	L	A	A	■	■
■	■	C	Y	R	■	■	D	E	S	C	R	Y	■	■
F	I	L	E	T	M	I	G	N	O	N	■	G	O	R
T	B	O	L	T	■	V	O	L	E	■	A	R	N	O
H	E	L	I	O	■	A	N	A	T	■	M	A	E	S
S	T	A	A	N	■	R	O	T	H	■	T	W	E	E

PUZZLE SOLUTION 95

G	A	S	P	■	C	O	N	G	■	D	C	L	I	I
E	L	O	I	■	H	C	H	I	■	R	E	A	D	E
S	A	M	A	R	I	T	A	N	■	O	N	D	E	R
S	I	M	■	I	R	E	■	N	A	P	T	I	M	E
O	N	E	A	C	A	T	■	I	D	I	E	■	■	■
■	■	E	E	C	■	L	E	B	■	R	H	E	T	■
S	N	O	O	D	■	S	E	M	I	■	S	U	E	R
S	E	E	L	■	D	T	O	A	Z	■	T	R	E	A
R	E	N	I	■	A	A	R	E	■	W	A	L	E	D
S	T	O	A	■	R	I	A	■	E	R	G	■	■	■
■	■	N	L	E	R	■	L	A	Y	E	R	E	D	■
P	E	T	H	A	I	R	■	E	R	L	■	U	S	O
O	C	E	A	N	■	A	N	O	N	Y	M	I	T	Y
O	R	A	R	E	■	M	O	V	E	■	O	N	E	O
P	U	M	P	S	■	P	A	I	R	■	E	S	S	U

PUZZLE SOLUTION 96

E	F	F	S	■	R	A	N	D	■	L	T	C	O	L
L	E	I	A	■	O	N	T	O	■	C	O	H	O	E
O	N	S	L	A	U	G	H	T	■	H	Y	E	N	A
I	N	H	A	L	E	R	S	■	S	A	B	R	A	S
■	■	D	K	N	Y	■	T	P	I	■	R	S	T	■
D	A	I	S	Y	■	■	W	H	A	M	M	Y	■	■
C	N	N	■	L	A	W	Y	E	R	■	O	B	I	E
I	N	T	D	■	G	O	A	P	E	■	L	O	N	G
I	S	E	T	■	E	L	T	O	R	O	■	M	C	A
■	■	S	S	H	I	F	T	■	D	A	B	L	L	■
A	S	T	■	A	N	E	■	A	G	E	S	■	■	■
S	K	I	I	N	G	■	P	S	A	L	T	E	R	S
C	A	N	A	L	■	T	U	T	E	L	A	G	E	S
I	L	E	N	E	■	D	L	I	L	■	G	O	S	S
I	D	S	A	Y	■	S	I	R	S	■	E	S	E	S

PUZZLE SOLUTION 97

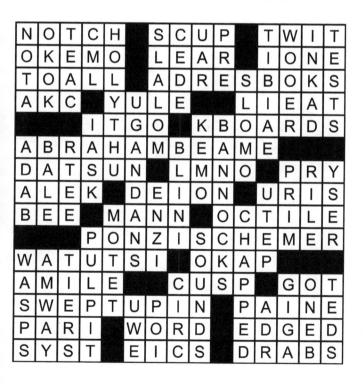

```
D I L L _ L I M O _ D E L H I
O N C E _ S N C C _ O N I O N
G R E E N A R C H I T E C T S
M U S _ I T E A _ S C R E W S
A N Z A C _ Y O H O _ _ _ _ _
_ _ D K N Y _ H E M A T I N
M O J O _ R A I L _ M E A D E
O P E N P A N D O R A S B O X
S I S I S _ K N O T _ O U S T
S A U S A G E _ K E L P _ _ _
_ _ _ L I E U _ _ I S S E T
K I B I T Z _ N O M O _ I R R
M A R I E A N T O I N E T T E
A N A I R _ D I M E _ S K E W
N A Z I S _ Y E E S _ S A S S
```

PUZZLE SOLUTION 98

```
Y Y Y Y _ H I L L S _ C O Z Y
O O M E _ A P A T H _ A N C E
U N C L E V A N Y A _ V E G S
I N A P I E _ A R M L O C K S
N E S _ D N C _ P A R _ _ _
_ _ G E T O U T O F T H E Y
R A N D R _ A L L O F _ U S O
A T A P _ A T N O S _ I S T H
M A V _ S H E A F _ E N E R O
P R I M R O S E P A T H _ _ _
_ _ R O Y _ Z A C _ S B C
H E L M S M A N _ A H C H O O
I S A O _ A Q U A M A R I N E
D E N T _ T H I S A _ U R D U
E L I O _ E S T O P _ S T I R
```

PUZZLE SOLUTION 99

```
N O T C H _ S C U P _ T W I T
O K E M O _ L E A R _ I O N E
T O A L L _ A D R E S B O K S
A K C _ Y U L E _ L I E A T
_ _ I T G O _ K B O A R D S
A B R A H A M B E A M E _ _
D A T S U N _ L M N O _ P R Y
A L E K _ D E I O N _ U R I S
B E E _ M A N N _ O C T I L E
_ _ P O N Z I S C H E M E R
W A T U T S I _ O K A P _ _
A M I L E _ C U S P _ G O T
S W E P T U P I N _ P A I N E
P A R I _ W O R D _ E D G E D
S Y S T _ E I C S _ D R A B S
```

PUZZLE SOLUTION 100

```
L A H L X _ T S P S _ T H A R
A R E A S _ E A R P _ I A L E
T A R Z A N T H E A P E M A N
I S M E _ C O L L _ A I M E D
_ _ D J I N _ A S O N _ _ _
S O R _ A S S E T S _ W O R M
E V E R I _ V E G _ I D I E
G A S O L I N E S T A T I O N
T R A P _ N E N _ E H U D S
B Y T E _ R E I N U R _ M E A
_ _ A R O D _ I L Y A _ _
A T I D E _ A S S N _ S P Y S
A T E O N E S H E A R T O U T
R E A P _ R A I I _ R H E M E
E N T E _ L P N S _ S E T A T
```

CPSIA information can be obtained
at www.ICGtesting.com
Printed in the USA
LVHW100152271218
601859LV00008B/144/P